STANDARD LOAN

Renew Books on PHONE-it: 01443 654456
Help Desk: 01443 482625
Media Services Reception: 01443 482610

Books are to be returned on or before the last date below

Treforest Learning Resources Centre
University of Glamorgan

WATER RESOURCES IN JORDAN

Evolving Policies for Development, the Environment, and Conflict Resolution

EDITED BY

MUNTHER J. HADDADIN

RESOURCES FOR THE FUTURE
WASHINGTON, DC, USA

Printed in the United States of America

No part of this publication may be reproduced by any means, whether electronic or mechanical, without written permission. Requests to photocopy items for classroom or other educational use should be sent to the Copyright Clearance Center, Inc., Suite 910, 222 Rosewood Drive, Danvers, MA 01923, USA (fax +1 978 646 8600; www.copyright.com). All other permissions requests should be sent directly to the publisher at the address below.

An RFF Press book
Published by Resources for the Future
1616 P Street NW
Washington, DC 20036–1400
USA
www.rffpress.org

Library of Congress Cataloging-in-Publication Data

Water resources in Jordan / edited by Munther J. Haddadin.
    p. cm. -- ("An RFF Press book.")
    ISBN 1-933115-32-7 (cloth : alk. paper)
    1. Water-supply—Government policy—Jordan. 2. Water resources development—Jordan. 3. Water conservation—Jordan. I. Haddadin, Munther J. II. Series.
    TD313.J6W38 2006
    363.6'1095695--dc22   2006006468

The paper in this book meets the guidelines for permanence and durability of the Committee on Production Guidelines for Book Longevity of the Council on Library Resources. This book was typeset by Peter Lindeman. It was copyedited by Joyce Bond. The cover was designed by Maggie Powell.

Cover photos, left to right: (1) View from hills overlooking the Dead Sea; (2) Prince El Hassan Bin Talal inaugurates a groundwater well at Risheh in Wadi Araba; (3) map of Jordan; (4) Munther J. Haddadin at the Hasa distribution pool in Ghor Safi, south of the Dead Sea; and (5) the Jordan River, at the north of Israel. Photos provided by Munther J. Haddadin and iStockphoto.

ISBN 1-933115-32-7 (cloth)

About Resources for the Future *and* RFF Press

RESOURCES FOR THE FUTURE (RFF) improves environmental and natural resource policymaking worldwide through independent social science research of the highest caliber. Founded in 1952, RFF pioneered the application of economics as a tool for developing more effective policy about the use and conservation of natural resources. Its scholars continue to employ social science methods to analyze critical issues concerning pollution control, energy policy, land and water use, hazardous waste, climate change, biodiversity, and the environmental challenges of developing countries.

RFF PRESS supports the mission of RFF by publishing book-length works that present a broad range of approaches to the study of natural resources and the environment. Its authors and editors include RFF staff, researchers from the larger academic and policy communities, and journalists. Audiences for publications by RFF Press include all of the participants in the policymaking process—scholars, the media, advocacy groups, NGOs, professionals in business and government, and the public.

ISSUES IN WATER RESOURCE POLICY

Ariel Dinar, World Bank, Series Editor

Books in the *Issues in Water Resource Policy* series are intended to be accessible to a broad range of scholars, practitioners, policymakers, and general readers. Each book focuses on critical issues in water policy at a specific subnational, national, or regional level, with the mission to draw upon and integrate the best scholarly and professional expertise concerning the physical, ecological, economic, institutional, political, legal, and social dimensions of water use. The interdisciplinary approach of the series, along with an emphasis on real world situations and on problems and challenges that recur globally, are intended to enhance our ability to apply the full body of knowledge that we have about water resources—at local, country, and international levels.

For comments and editorial inquiries about the series, please contact *waterpolicy@rff.org*.

For information about other titles in the series, please visit www.RFFPress.org/water.

The *Issues in Water Resource Policy* series is dedicated to the memory of Guy LeMoigne, a founding member of the Advisory Committee.

Contents

Foreword

Jordan faces water poverty along with poverty in energy resources and financial assets. The management of water resources is a primary duty that the successive governments of Jordan have paid attention to. During my second tenure as prime minister, the cabinet, for the first time in the history of the country, adopted a water strategy, under which a series of policies were formulated. The editor of this book was the minister of water and irrigation, who prepared the drafts, presented them to the cabinet, and defended their contents before they were approved and issued as government water strategy and the succeeding policy on water utility, irrigation water, groundwater management, and wastewater management. Furthermore, he produced an investment plan for the water sector covering the period 1998–2010 and made it revolving in nature such that new projects can be added to expand the planning horizon and completed projects can be dropped from the plan. It has been the basis of water planning and implementation, with minor adjustments dictated by changing circumstances.

Prior to my assumption of the responsibilities of prime minister in May 1993, the editor of this book worked under me as a senior member of the Jordan delegation to the bilateral peace talks, where he was influential in shaping Jordan's policy toward negotiations on shared water resources. He competently led the water negotiations to a successful conclusion, ending on the morning of October 17, 1994. Concurrent to his work in the bilateral negotiations, he led the Jordan delegation to the multilateral Working Group on Water Resources, where he presented many ideas for regional cooperation on water resources. His experience in the development of Jordan's water resources, legislation for it, and overseeing the management of operation and maintenance of its projects qualified him to be an authority on the subject in the country. In short, there is hardly a professional who can more eloquently address the issues of water resource policy and implementation in the Hashemite Kingdom of Jordan than Munther J. Haddadin, who edited and contributed to this book.

The other authors whose work appears in this volume are recognized experts and practitioners in their fields. They occupy academic and professional positions in the government, the private sector, and nongovernmental organizations and have aided at various times in government efforts to develop the water sector. I am acquainted with most of the Jordanian contributors and am impressed with their competent additions to this volume.

The diversity of issues addressed in the various chapters of this book cast light on the water situation in the kingdom. The presentations and analyses make this a comprehensive reference for workers, researchers, and interested parties. The authors introduce innovative analytical approaches to the assessment of the total potential of water resources and review their social, economic, environmental, and political impacts. They present approaches for the first time to quantify soil water, evaporation rate and amount, and the cost of shadow water (virtual water) as compared with the cost in Jordan of irrigation water. Account is made of the peaceful settlement of the water conflict with neighboring Israel, in which the editor was a key partner.

The book is a welcome effort in its documentation of the policy aspects and their roots, and in its appraisals of them along with recommendations for improvement. Its discussion of water governance is of particular interest to water managers, policymakers, and other parties. It is the first such collective effort in Jordan, and the authors, editor, and publisher are to be thanked and congratulated for it.

Dr. Abdul Salam Majali
Former Prime Minister and
Current Deputy Speaker of the Senate
of the Hashemite Kingdom of Jordan

Figures and Tables

Contributors

Hani A. Al Rashid is a freelance irrigation consultant since 2000, before which he was a senior irrigation advisor with the irrigation department at the Ministry of Water and Irrigation in Jordan and its predecessor, the Natural Resources Authority. He has assumed leading roles in donor-funded projects (USAID, the Japan International Cooperation Agency, the Food and Agriculture Organization, and the Arab Fund for Economic and Social Development) addressing irrigation issues, and established a database on the influence of treated wastewater on irrigated agriculture.

Ra'ed Daoud is managing director of ECO Consult firm in Jordan. He has over 15 years of experience in water utilities management, water resource planning, regulatory reform, public private partnership for water utilities, and costing and pricing of water and wastewater services. Mr. Daoud participated in consultancy assignments for governments and international technical cooperation agencies, and facilitated discussions and meetings on water sector reform in Jordan and Middle East countries.

Franklin M. Fisher is the Jane Berkowitz Carlton and Dennis William Carlton Professor of Microeconomics, Emeritus, at the Massachusetts Institute of Technology, where he taught for 44 years. He is also chair of the Water Economics Project and senior author of its book, *Liquid Assets: An Economic Approach for Water Management and Conflict Resolution in the Middle East and Beyond*. His other 16 books include *The Identification Problem in Econometrics* and *Disequilibrium Foundations of Equilibrium Economics*.

Munther J. Haddadin is consultant and courtesy professor at Oregon State University and affiliate professor at the University of Oklahoma. He served Jordan as senior member of the Middle East Peace Process delegations and as Minister of Water and Irrigation. Dr. Haddadin's research focuses on the role of trade in alleviating water shortage and on the quantification of green water to be listed in the stock of water resources. He is author of *Diplomacy on the Jordan*, and coauthor of *Liquid Assets: An Economic Approach for Water Management and Conflict Resolution in the Middle East and Beyond* and *Peacemaking: The Inside Story*. He has been decorated for excellence by King Hussein of Jordan, Queen Elizabeth II, the presidents of Germany, Austria, and Italy, and the Patriarch of Jerusalem, and is the recipient of several prestigious scientific awards.

Emad Karablieh is associate professor of agricultural economics and agribusiness at the University of Jordan. His academic interests include water economics, risk analysis, and integrated water resource management. His professional consultations cover environmental economics, water and agricultural policies, technology adoption, and investment appraisal. His publications appeared in *Agricultural Water Management, Agricultural Systems, Journal of Arid Environment,* and *Quarterly Journal of International Agriculture.*

Hala Zawati Katkhuda is proprietor and manager of EasyInfo consulting firm. She is an expert in information systems management, and specialized in natural resources and environment data acquisition, compilation, processing and management. She contributed to numerous projects to establish and enhance water information systems in Jordan and Mediterranean countries. She worked as a consultant for the United Nations Development Programme, the European Union, USAID, GTZ, and the Japan International Cooperation Agency, and is a 2005 Jordanian Eisenhower fellow.

Helena Naber is an environmental analyst with the United Nations Development Programme in Jordan. Before joining UNDP, Ms. Naber was an environmental consultant with ECO Consult. Her work experience and interests include a wide array of environmental issues such as UN environmental conventions (CBD, UNFCCC, and CCD), environmental impact assessment, water resources economics and management, and environmental valuation.

Razan Quossous is water treatment utilities consultant with ECO Consult. She has been involved in and has managed projects related to sustainable development, water resources management, wastewater treatment and reuse, social assessment related to water use, infrastructure development, private sector participation, and public utilities support, and has also conducted environmental assessments related to wastewater issues.

Elias Salameh is professor of hydrogeology and hydrochemistry at the University of Jordan and founder of its Water Research and Study Center. He is associate editor of *Hydrogeology Journal, Hydrogeologie und Umwelt,* and *Acta hydrochimica et hydrobiologica,* and has published over 140 papers in international journals on hydrogeology, hydrochemistry, water resources management, and water resources in the Middle East. In 2004 and 2005 he chaired the founding committee of the German Jordanian University.

Amer Salman is an associate professor in the Department of Agricultural Economics and Agribusiness at the University of Jordan, where he is active in teaching, research, supervising graduate students, and professional consultation. His topics of interest include economics of water resources and use with a focus on quality and pricing, and on environmental economics of natural resources. His publications appeared in *Water International, Agricultural Water Management, Agricultural Systems,* and *Quarterly Journal of International Agriculture.*

Musa Shteiwi is associate professor of sociology at the University of Jordan and the founder and director of the Jordan Center for Social Research. He is author of *Al-Tatawae: Volunteering in the Arab World,* and Arabic publications: *Social Policies in Jordan, Gender Roles Stereotyping in Primary School Textbooks in Jordan, Class Structure of Jordanian Society,* and coeditor of *New Media and Youth in the Arab World.*

Sami J. Sunna' is director of MEMEARE consulting firm, and advisor to the Jordanian Government on the implementation of the National Strategy for Agricultural Development, in the preparation of which he participated. He held the positions of director general at the Agricultural Credit Corporation and secretary general at the Ministry of Agriculture in Jordan, was director of the joint FAO-ESCWA Agricultural Division, and a member of the International Service for National Agricultural Research (ISNAR) board of trustees.

Mai Abu Tarbush is an environment and wastewater consultant with a USAID-funded project for the Reuse of Wastewater for Industry, Agriculture and Landscaping, implemented by CDM International. She previously worked as a consultant with ECO Consult, and has broad experience in industrial wastewater treatment, wastewater management and reuse, environmental due diligence, pollution prevention and control, and implementation of environmental management systems.

Heinz-Peter Wolff is assistant professor in agricultural and social economics at the University of Hohenheim, Germany, and a former natural resource economist at the International Centre for Agricultural Research in the Dry Areas. He cooperates with colleagues from the University of Jordan in a joint research program on integrated water resource management and supervises, among others, the Ph.D research of Jordanian students in this field.

Introduction

Munther J. Haddadin

Water Resources in Jordan: Evolving Policies for Development, the Environment, and Conflict Resolution is the first book to tackle this important subject and its impact on the socioeconomic development of the country. It follows the first attempt ever undertaken in the country by the minister of water and irrigation to formulate a government strategy on water resources, endorsed by the Council of Ministers in 1997.

The book is organized into 11 chapters plus this Introduction, which describes the chapters and their interconnection and highlights the book's innovative elements. The chapters discuss policy issues in connection with the specific material covered in each, make recommendations to improve or add to policies where necessary, and touch briefly on the interactive aspects of water resources with some countries of the region.

Chapter 1 examines the water resources of Jordan (see Figure I-1 for a map of the country), including their occurrence, average flows of surface water, and safe yields of groundwater reservoirs. It quantifies the indigenous resources from rainfall over the kingdom, as well as those originating in rainfall outside Jordan. The average evaporation rate is calculated and compared with the world average.

The chapter also assesses the adequacy of water resources to meet the demand and shows the distortion in the population–water resources equation. All forms of water resources are considered: blue water (groundwater and surface water), green water (soil water), silver water (desalinated seawater), and gray water (treated wastewater). The significance of commodity imports and their role in balancing the population–water resources equation become clear. Innovative approaches are used in the calculation of the evaporation rate and the quantitative definition of the impact of overabstraction from groundwater aquifers on spring flow (base flow) and, separately, on the groundwater table.

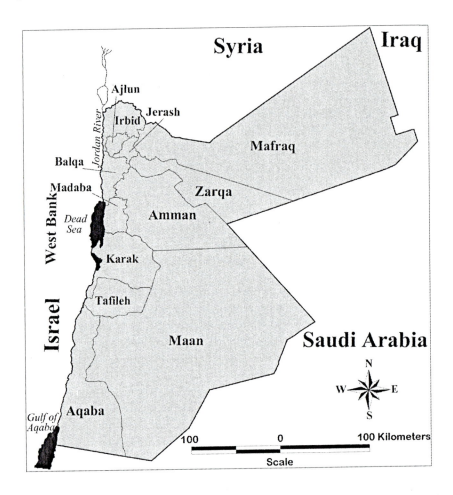

FIGURE I-1. Map of Jordan
Source: MWI 2005.

　　Chapter 2 reviews the history of water legislation in Jordan from as far back as 1937, when the country was under the British Mandate. The "oneness" of water legislation and its institutional setup in Jordan is highlighted. The chapter examines the diversity of approaches to water management, as reflected by the corresponding legislation, and looks at the latest arrangement, building on the country's long experience and the imbalances in the population–water resources equation. Weaknesses in some pieces of legislation are pointed out, and recommendations are made to rectify them. The chapter concludes with a look ahead to future legislation needs for the management of water resources.

　　Water data management is addressed in Chapter 3. It reviews the country's past experiences and describes the advanced data systems Jordan now is using thanks to assistance from friendly donors. The need for transparency in data acquisition and management is stressed, and the reasons for hiding data away from public circulation are examined. Recommendations are made to expedite Jordan's involvement in the modern data systems and databanking.

In Chapter 4, efforts to develop Jordan's water resources and expand and improve irrigation systems are discussed. The chapter looks at municipal supplies and the imperative of remote interbasin transfers, competition among user sectors, the involvement of government in irrigation development and its approach of integrated planning and implementation, and how the introduction of modern systems and practices in the Jordan Valley resulted in improved irrigation efficiency and crop yields.

Chapter 5 deals with environmental issues related to water resources. Policies in regard to the environment and the allocation of environmental water are covered. The chapter also examines the impact of ignoring water for the environment on several locations in the kingdom.

The economics of water in Jordan are addressed in Chapter 6, including the real cost of water, tariffs, and subsidies. The Jordan water sector is shown to lack necessary prerequisites to become a free-market system. The chapter presents the Water Allocation System model, which serves as a good tool the government can use in the allocation of water resources and development planning.

Chapter 7 examines the role of trade in dealing with the water shortage, looking at how it helps Jordan bridge the wide gap between available supplies of water and demand. The pressure caused by the population's demand for water is termed "water stress," and the impact of water stress on society is called "water strain"; the rate of change of stress with respect to strain is known as the "water modulus." The higher the water modulus, the higher the water stress and water strain. A new term, "shadow water," is introduced to refer to the indigenous water saved through the import of industrial and agricultural commodities, and the use of the term is justified. The water threshold is considered in the context of an imagined "virtual environment." An innovative analytical approach to the assessment of green water is attempted, thus allowing for the quantification of all the water resources available to the country: blue, green, silver, and gray.

The challenges facing the water sector in Jordan are presented in Chapter 8. Addressed are issues of incremental supplies; unsustainable use of groundwater aquifers; water allocation and reallocation; human resource development; the cost of energy, all of which is imported; water governance, including stakeholder participation; equity; affordability of escalating costs; enabling the poor; and the dilemma of cost recovery. The chapter also examines the recent trend of government partnership with the private sector and the transfer of administration to companies. Government policy in regard to these issues is outlined and discussed. Recommendations are made for future management of blue, green, and gray water.

Chapter 9 looks at the linkages between water resources and social matters, including the impact that access to water networks has on women, young people, and the poor, as well as the impact of water rationing on the poor and their crops. Some inequity issues in water allocation for urban areas and irrigation are mitigated by subsidies accorded to the less well-to-do. Bias in water allocation toward urban uses is mitigated by the policy of treated wastewater reuse in irrigation and by subsidies accorded to irrigation water and agriculture in general. The case of the integrated development of the Jordan Valley, with irrigated agriculture as its backbone, is presented to demonstrate the social and cultural impacts water has on societies. The government policy of adopting the integrated-development approach yielded good results in the economic and social spheres in the Jordan Valley. Drastic improvements were seen in education, particularly for females, as well as in health conditions and care, with better provision of preventive and curative measures, declines in child morbidity and mortality, and increased life expectancy at birth. Impact on the quality of life was positive, and the valley residents' participation in the political life of the country became visible. Cultural transformations paralleled the social changes.

Chapter 10 addresses the issues related to international water bodies on which Jordan is a riparian party. It includes a detailed discussion of the Jordan Basin, elaborating on the conflict with Israel and its peaceful resolution in the 1994 peace treaty between Jordan and Israel. A decrease in Jordan's shares in the Yarmouk River resulted from Syrian abstraction from the surface-water and groundwater resources that had been allocated to Jordan in the first treaty between the two countries in 1953. The details of the settlement of the conflict with Israel are presented, with reference to the Palestinian dimension. Shared groundwater resources, mostly fossil water, are pointed out. The chapter highlights the need to revisit the treaty with Syria and the implementation of the agreement with Israel to conform to the provisions of the Water Annex in the 1994 treaty. It stresses the necessity to legislate for the implementation of policy regarding Jordan's international waters.

The final chapter, Chapter 11, draws conclusions from the presentations in previous chapters of water resource availability and government policy. It reviews qualitative and quantitative concerns, including the erosion of Jordan's shares from international waters, and recommends actions the government should take to address them. Because of the rising importance of treated municipal wastewater as a resource, it demands close attention for better management, and the significance of shadow water necessitates focus on trade policy and foreign policy with respect to countries that are trading partners.

Where appropriate, the authors have attributed data and information to their sources. Any unreferenced data have been taken from the files of the Ministry of Water and Irrigation. Widely known general information is not referenced; all other unreferenced information is the product of the authors' analyses.

The local currency, the Jordan dinar (JD), is occasionally used in the economic and financial analyses, because the source of information on costs, revenues, and prices is local. The local currency has been tied to the U.S. dollar since the early 1990s, and the current conversion rate is US$1.41 to the Jordan dinar.

Water resources in most Arab and Islamic countries are considered public property and have been managed by government agencies or municipal councils in line with this outlook. In many countries of the region, public property extends to all natural resources occurring on the surface or in the subsurface. Government policies with respect to water resources are ordinarily drafted by the concerned government agency in charge, usually in collaboration with other government agencies or ministries directly affected by these policies. The outcome of such consultations is usually reflected in the draft, and the policy is formulated and approved by the Council of Ministers. The need to submit the draft policy to the Council of Ministers is dictated by the fact that water resources touch on the economic, social, and environmental activities of all sectors of the economy and have direct impact on the well-being of the people.

In their water allocation policies, practically all countries of the Middle East, including Jordan, accord first priority to municipal and industrial uses. Their industrial uses fall below those in developed countries. Agricultural uses occupy second place. Environmental uses have been properly valued only recently, and in some, not all, countries of the region.

In none of the Middle East countries has the administration of water resources and their uses been made the responsibility of one entity. In Jordan, for example, the Ministry of Water and Irrigation has charge of water resources, with the exception of green (soil) water that supports rain-fed agriculture; this falls within the jurisdiction of the Ministry of Agriculture as the government agency and the farmers as the operators of dry-land farming. In Saudi Arabia, the Ministry of Water and Agriculture has charge of water for irrigation and caters for rain-fed agriculture, but municipal and industrial water, drawn from natural sources, is the responsibility of the Ministry of Municipal Affairs; silver water (desalinated seawater) carries much weight in the municipal and industrial supplies in Saudi Arabia and is the responsibility of agencies independent from these two ministries. Egypt, Syria, and Iraq each have a Ministry of Irrigation that carries the burden of irrigation infrastructure and its operations, as well as a Ministry of Agriculture that cares for the on-farm activities of both irrigated and rain-fed agriculture; municipal and industrial supplies are the responsibility of the Ministry of Municipal and Rural Affairs. In all these countries, farmers' associations and cooperatives play an important role in providing agricultural inputs and influence marketing of the produce. Government has a role to play in some form or another.

Other Gulf states have Ministries of Water and Energy; these are countries where irrigation does not claim a substantial portion of their water resources, and where silver water commands a substantial role in the supply of municipal and industrial water. In Israel, a Water Commission, whose head reports to the minister of national infrastructure, formulates and executes the country's water policy. Infrastructure is the responsibility of a government company, Mekorot, and water studies are undertaken by yet another quasi-government agency, Tahal. Green water is the responsibility of the Ministry of Agriculture and the agricultural cooperatives, the moshavs.

In all the countries of the region, irrigation water is subsidized, as are municipal and industrial water. In oil-rich countries, the subsidy is greater, and so is the cost of water service. In Saudi Arabia, for example, water is desalinated on the Red Sea coast in Tihama, and the silver water is pumped over a very high head, believed to be the highest in the world, to supply municipal water to Abha, Khamis Msheit, and environs. The charges for municipal water hardly cover a small fraction of the high cost.

For nearly 40 years, Jordan associated its irrigation water policy with measures aimed at irrigated land reform. In all government-sponsored irrigation projects, a land reform component is embedded. That policy succeeded in achieving social development and enabling the poor farmers. It was later associated with provisions prohibiting the free sale of irrigated farm units, but it allowed such sale, at predetermined prices specified in the law, to the Jordan Valley Authority, the government arm in the development of the Jordan Valley. The owners of the lands before irrigation received an equitable share of their unproductive land, and the excess was redistributed to landless farmers or owners of smaller holdings. In neighboring countries, especially in Egypt, Syria, and Iraq, land reforms were biased against landowners, and the economic and social objectives were not met as successfully as they were in Jordan.

Whereas water laws in the countries of the region have more or less similar roots and philosophy, the administration of water resources has varied at different stages, but most of the countries went through similar experiences of decentralization and centralization of water administration. Only Jordan adopted the approach of integrated social and economic rural development, with water, through irrigation, acting as a backbone. Water stress and the associated water strains differ from one country to another, and that is reason enough for countries to have different approaches in water management.

Reference

MWI (Ministry of Water and Irrigation). 2005. *Water Management Information System.* Amman, Jordan: MWI.

1

The Population–
Water Resources Equation

Elias Salameh and Munther J. Haddadin

This chapter examines the balance between Jordan's population and available water resources. It begins by looking at the kingdom's population and demography, then reviews the topography, climatic zones and associated precipitation, evaporation and evapotranspiration, and surface-water and groundwater basins with their indigenous and exogenous renewable yields. The precipitation balance among surface runoff, groundwater recharge, greenwater retention, and evaporation is addressed, with loss to evaporation calculated at about 80% of precipitation. Overabstraction from renewable aquifers and the use of fossil, nonrenewable aquifers are assessed. The chapter ends with projections of future water demand and the prospects of meeting this demand.

The Population

The population profile of Jordan has been impacted by the political developments in the Middle East. At its infancy in 1922, the Emirate of Transjordan had a population of 225,000 people, which increased to 300,214 by 1938 and 473,222 by 1947. By early 1948, the influx of Palestinian refugees commenced, and by 1951, the population had jumped to 599,562 people. The population in 2002 amounted to about 10 times that in 1949, a tenfold increase over about half a century. The natural population growth was capable of doubling the population every 22 years on average, which would have increased the original population level, without the influx of refugees, to 2.84 million in 2004 compared with the actual 5.35 million it reached that year (DOS 2004).[1]

The above figures do not include the population of the West Bank, an integral part of Jordan between 1950 and 1988, when the king issued a decree of legal and administrative disengagement with the region. It had fallen to Israeli occupation in the June war of 1967. Its liberation from occupation became the concern of the Palestine Liberation Organization (PLO). The disengagement decree created a fifth riparian party on the Jordan basin, the Palestinians, in addition to Lebanon, Syria, Israel, and Jordan.

The demographic composition of the population of Jordan consists mainly of original Jordanians (East Bankers), Palestinians (West Bankers and refugees), Syrians, and people in other, smaller communities with origins from Hijaz, Iraq, and the Caucasus. This demographic mix affects the policies of the kingdom in regard to its stand toward peace in the Middle East and potential water relations with its neighbors, particularly Syria and the Palestinian Authority.

The Water Resources

The kingdom's renewable freshwater resources originate in precipitation over its territories and the flows of international watercourses to which Jordan is a riparian party. These international watercourses are the Jordan River and its largest tributary, the Yarmouk.

The topography of the country and its vegetative cover affect the rates of precipitation. The rainy season extends from October to April, peaking during January and February of each year. Precipitation in Jordan is mainly in the form of rainfall. Snowfall generally occurs once or twice a year over the Highlands, a relatively narrow strip of plateau running north–south, paralleling the Jordan Rift Valley.

Geographically, the country has three distinct zones. The first is the Jordan Rift Valley, a north–south strip lying primarily below sea level. This strip contains the Jordan River Valley, which slopes gently from north to south where the river discharges into the Dead Sea. The shoreline of the Dead Sea is the lowest dry contour on earth, at 418.30 meters below sea level in 2006. It has three riparian parties: Jordan encompasses half of its surface area, and Israel and the Palestinian Authority each roughly a quarter. The rift continues through the Southern Ghors and Wadi Araba, mostly a semidesert terrain rising gently from the Southern Ghors to Risha at 230 meters above sea level, and then slopes gently toward sea level at the Gulf of Aqaba. The second is the Highlands, or the Jordanian Plateau, a north–south strip containing the southward extension of the Anti-Lebanon mountain range, traversed by the Yarmouk River, the largest tributary of the Jordan, and several other side wadis of perennial flow discharging into the Jordan River; most of Jordan's cities are located in this strip. The third is the Badia, a semidesert steppe forming the rest of the territories of the country to the east, north, and south.

The highest rates of precipitation fall over the Highlands strip, and the long-term annual averages decrease from north to south, with 600 mm in Ajloun,

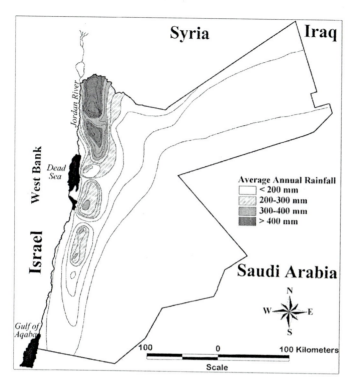

FIGURE 1-1. Average Rainfall in an Average Year
Source: MWI 2005a.

550 mm in Balqa, 350 mm in Karak, and 300 mm in Shoubak. Precipitation decreases to the east of the Highlands and even more sharply to the west; it decreases, for example, from an average of 600 mm/y in Ajloun to 250 mm/y in the Jordan Valley within a distance of 10 km due west and a drop in elevation of 1,200 m. The decrease in the westerly direction is two to three times greater than is the case in the easterly direction: from 300 mm/y in Rajef, for example, to 50 mm/y some 30 km east of the town. Figure 1-1 shows the precipitation distribution over Jordan in an average year.

The general characteristics of precipitation and its budget categories can be summarized as follows:

- The average annual amount of water falling over Jordan's territories is about 8.2 bcm, increasing to 12 bcm in a wet year and decreasing to 6 bcm in a dry year.
- Only a minute portion of Jordan's territories, about 1.3%, receives an average annual precipitation of more than 500 mm; 1.8% receives between 300 and 500 mm; 3.8% between 200 and 300 mm; 12.5% between 100 and 200 mm; and the majority, 80.6%, less than 100 mm.

- The climatic conditions not only affect the amount and distribution of precipitation, but also are a decisive factor in evaporation and evapotranspiration potentials. The potential evaporation rates range from about 1,600 mm/y in the northern Highlands to more than 4,000 mm/y in the southern desert and the Badia territories at Azraq, about 75 km to the northeast of Amman. These rates are 5 to 80 times the corresponding average amounts of annual precipitation over their respective areas. Evaporation, as will be shown below, accounts for about 80.2% of the precipitation over the territories of the kingdom.
- The dry climate, atmospheric dust, and low intensity of precipitation affect the quality of precipitation water, generally reflected in higher salt contents.

The above helps paint a clear picture of Jordan's water situation; only about 3% of the total area of the country receives an average annual intensity of precipitation exceeding 300 mm, the least amount needed to grow wheat in rain-fed farming areas under the prevailing climatic conditions of the country. Such areas receive about 17% of the country's annual precipitation, while the majority of precipitation, or 83%, occurs over other areas with lower rates of precipitation, where rangelands may be supported by 150 to 300 mm/y (Qasim 1993).

As is the case elsewhere in the world, Jordan's precipitation balance falls into four categories: The first, with a short residence time, is lost to evaporation, yielding practically no economic benefit. The second is surface runoff, which forms part of the water cycle as the water returns to the bodies where it originated. The third infiltrates down to the aquifers and forms the groundwater resources of the country. The fourth is soil water, which coats the soil particles through surface tension and is held between the particles by capillary action; this water supports dry-land farming, rangelands, forests, wetlands, and natural vegetation. The second and third categories together constitute what is termed "blue water," and the fourth category is termed "green water."

Surface Drainage Basins

Jordan has seven distinct surface drainage areas (Figure 1-2). Most important among them is the Dead Sea Basin, by virtue of its quantities of usable water flows, the percentage of population resident in it, and the non-Jordanian territories it drains in the neighboring countries of Lebanon, Syria, Israel, and the Palestinian territories. Topography indicates that a small portion of Egyptian territories in the Sinai would drain into the Dead Sea via Wadi Araba if surface runoff were enough to reach it. Several basins shown in the figure belong to this major basin of the Dead Sea (seven in one basin). These are the Yarmouk, the Jordan River and side wadis, the Zarqa River, Dead Sea side wadis, Mujib Basin, Wadi Hasa, and Wadi Araba North.

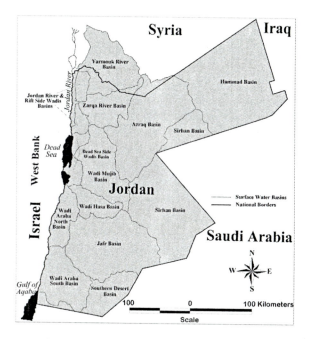

FIGURE 1-2. Surface-Water Basins
Source: MWI 2005a.

The Jordan River

The Jordan River's terminus is the Dead Sea. It starts at the confluence of three upper tributaries: the Hasbani, which drains territories in Lebanon and Syria; the Dan, which begins inside Israel and drains territories at the foothills of Mount Hermon; and the Banyas, which drains territories in the northwest of Syria's Golan Heights. The Jordan flows southward, entering and exiting Lake Tiberias before it is joined by its major tributary, the Yarmouk, which drains territories in Syria and Jordan and forms the borders between Jordan and Israel shortly before its confluence with the Jordan. Several other side wadis discharge into the Jordan from the east and west before it reaches its terminus, the Dead Sea. As such, the Jordan Basin is an international water basin.

Different estimates were made of the river discharge into the Dead Sea. In the absence of hydraulic measurements, an estimate of 1,370 mcm/y was made by the U.S. consultants Chas T. Main in 1953 for use in a hurriedly formulated plan for the development of the Jordan Valley (Haddadin 2001). Another estimate of the Jordan Basin total discharge amounted to 1,348 mcm (Haddadin 2001). But the estimate closer to reality, 1,213 mcm/y, was made in 1954 based on data from flow gauges installed on the Jordan and its tributaries and through analysis of historic rainfall data collected from monasteries in Jerusalem (Baker and Harza 1955). Interestingly, this total flow of the Jordan River, important to five riparian parties, is approximately equal to 1.5% of the annual flow of the Nile in Africa at Aswan in Egypt, or 4.3% of the Euphrates River flow at Qai'm in Iraq.

The Yarmouk is the international watercourse that concerns Jordan. The riparian parties on it are Jordan, Syria, the Palestinians, and Israel. Contributions to its flow come primarily from springs emerging in its catchment in Syria and from surface runoff of about 5,630 km^2 in Syrian territories and 1,160 km^2 in Jordanian territories. The natural annual flow of the river was calculated at 467 mcm, increased by virtue of irrigation return flow from Syria (39 mcm) and reduced by the estimated evaporation (14 mcm) to arrive at a figure of 492 mcm.[2]

The next important to Jordan among the tributaries is the Zarqa, draining about 3,225 km^2 extending as far north as the foothills of Jebel Druz inside Syria, with an average natural flow amounting to 92 mcm/y (Baker and Harza 1955). The natural state of the river was altered by different factors, most notable of which were the drying up of most of its freshwater base flow as a result of overpumping from the aquifer that feeds its springs to supply the expanding capital city of Amman; the construction on it of the King Talal Dam in 1977 and raising it in 1987; importing water into the basin through several projects of interbasin transfers for municipal and industrial uses; and building wastewater treatment plants and discharging their effluents into the river system. In addition to affecting the river flow and its regulation, these activities had adverse impacts on the environment and on water quality.

Other Jordanian tributaries to the Jordan River, flowing from east to west and unshared by other riparian parties, are, from north to south, Wadi Arab (regulated by a dam since 1987), Wadi Jurum, Wadi Ziglab (regulated by a small dam since 1963), Wadi Yabis, Wadi Rajib, Wadi Kufrinja, Wadi Shueib (regulated by a recharge dam since 1966), Wadi Kafrein (regulated by a modest dam since 1968), and Wadi Hisban. Their total average annual flow combined with the Zarqa River was 213 mcm (Baker and Harza 1955), of which 175 mcm were considered usable (Haddadin 2001). The Jordanian share in the Jordan Basin averaged about 471 mcm/y (Chapter 10).

Southern Dead Sea Basin

South of the Jordan Basin, but within the Dead Sea Basin, are streams discharging directly into the Dead Sea. The primary ones are Mujib and Hasa. The annual flow of all the wadis combined averages about 180 mcm. A modest additional contribution is made from Wadi Araba north, which drains about 5.5 mcm into the Dead Sea.

Other Basins

There are six other surface-water basins: the Red Sea (including southern Wadi Araba), Southern Desert, Jafr, Sirhan, Hamad, and Azraq. Although their catchments are sizable, their water contributions are modest, as shown in Table 1-1.

Exogenous Surface Water

Jordan's share in the Yarmouk has been diminished by excessive Syrian withdrawal from its catchment. In 1992, about 61.2 mcm of base flow and 122.6 of flood flow, a total of 183.8 mcm/y on average, came from the Syrian part of the catchment, while the contributions of the Jordanian territories to the Yarmouk were a total of only 41.2 mcm, in the form of 18.8 mcm of base flow and 22.4 mcm of flood flow.

The 1994 Jordan–Israeli Peace Treaty (see Chapter 10) guaranteed a transfer of 100 mcm/y of water to Jordan, broken down into 35 mcm from joint development of the Lower Jordan, of which about 20 mcm would come from spills of Lake Tiberias; 60 mcm from desalinated water (as of 2004, only 35 mcm had been delivered on a regular basis); and 20 mcm from Lake Tiberias as a swap with 20 mcm of Jordanian floodwater. The exogenous resources of surface water totaled about 283.8 mcm, and the indigenous resources totaled 586.8 mcm/y, consisting of 229.9 base flow and 356.9 mcm flood flow. However, a portion of the base flow emerges from a fossil aquifer in the amount of 33 mcm/y; hence the renewable base flow is adjusted to 197 mcm/y, and the total indigenous surface flow would thus amount to 554 mcm/y.

Groundwater Aquifers and General Flows

Jordan has 12 groundwater reservoirs, with aquifers ranging in formations consisting of carbonates, basalts, and sandstones. They are divided into three main categories: Deep Sandstone Aquifer Complex, Upper Cretaceous Aquifer Complex, and Shallow Aquifer Complex.

Deep Sandstone Aquifer Complex

The Deep Sandstone Aquifer Complex forms one unit in southern Jordan. To the north, gradually thick limestone and marls separate it into two aquifer systems, which nonetheless remain hydraulically interconnected. Rocks forming this complex consist of Disi Group Aquifer (Paleozoic) and Kurnub and Zarqa Group (Lower Cretaceous Age).

Upper Cretaceous Aquifer Complex

The Upper Cretaceous Aquifer Complex consists of an alternating sequence of limestone, dolomite, marlstone, and chert beds. The total thickness in central Jordan is about 700 m. The limestone and dolomite units form excellent

TABLE 1-1. Summary of Surface Water Resources (mcm/y)

Basin	Base flow	Flood flow	Total flow	Used portion (2002)
Jordan River below Tiberias				
Jordan's share	15[a]	20[b]	35	0
Transfers from Israel	60[c]	—	60	35
Yarmouk River	$(18.8 + 20^d$ $+ 61.2^e) = 100$	$(22.4 + 122.6^e)$ $= 145$	$(41.2 + 203.8)$ $= 245$	145
Zarqa River	52.6	47.4	100	100
Side wadis north	31.8	18	49.8	45
Side wadis south	29.4	25.1	54.5	50
Dead Sea side wadis	22.6	21.7	44.3	17
Wadi Mujib	35	70.9	105.9	0
Wadi Hasa	28	13.1	41.1	30
Wadi Araba North	11.7	34.3	46	12
Wadi Araba South	0	7.8	7.8	0
Subtotal	$(229.9 + 15 +$ $81.2 + 60)$ $= 386.1$	$(260.7 + 20 +$ $122.6) = 403.3$	$(490.6 + 203.8 +$ $95) = 789.4$	434
Azraq (Jordan)	0	40.9	40.9	0
Jafr	0	12.5	12.5	0
Hamad	0	24.3	24.3	0
Sirhan	0	17.5	17.5	0
Southern Desert	0	1	1	0
Total	$(229.9 + 15 +$ $141.2) = 386.1$	$(356.9 + 142.6)$ $= 499.5$	$(586.8 + 283.8)$ $= 885.6$	434

Notes: Total surface water generated inside Jordan = 229.9 base flow + 356.9 flood flow = 586.8 mcm/y. The reuse of a flow of 15 mcm of agricultural return water is discounted. In the base flow, there is a portion of fossil water that calls for a correction of the total base flow generated by renewable water to be 197 mcm. The total runoff from rainfall on the territories of the kingdom is corrected to be 554 mcm/y on average. Also, the base-flow regime is distorted in the field because of overabstraction from groundwater reservoirs feeding the watercourses with base flow. Compared with the sustainable yield of wells of 266 mcm/y, the total actual abstraction in 2002 was 433 mcm, overshooting the sustainable yield by 167 mcm.

a. Return flow from irrigation in northern Jordan Valley, quantity part of 35 mcm implied in peace treaty.

b. Flood flow share in the Jordan River coming from Tiberias spills and other uncontrolled side wadis.

c. Flow coming from Israel as per peace treaty.

d. Flow in return for Jordan's concession to Israel to pump 20 mcm from Yarmouk floods.

e. Flow coming from Syria.

Source: ROID 1993.

aquifers. The lowest portions of this sequence, consisting of about 200 m of marls and limestone, possess relatively high permeability in some areas and form a potential aquifer. An aquitard consisting of about 80 m of marl and shale overlies the first sequence and separates it from the overlying aquifers.

Shallow Aquifer Complex

The Shallow Aquifer Complex consists of two main systems: the Basalt Aquifer and the Sedimentary Rocks and Alluvial Deposits of Tertiary and Quaternary ages.

Groundwater Basins

Groundwater basins, or groundwater balance areas, are those areas that could be demarcated and separately defined to include appropriate and regionally important aquifer systems. The groundwater divides are either aquifer limits or important and relevant geomorphologic or geologic features.

Groundwater basins in Jordan (Figure 1-3) are demarcated according to the same criteria, with some of these basins' recharge and discharge taking place within the same basin. In most basins, more than one aquifer complex is present, and hence any definition of groundwater basins should refer to a certain aquifer system and not necessarily to all aquifer systems underlying the basins.

The *National Water Master Plan* of Jordan (GTZ 1977) made such definition of the groundwater basins in Jordan and enumerated them as 12 distinct basins. In this work, the Jafr Basin in that plan is subdivided into Jafr and Disi-Mudawwara, raising to 13 the total number of groundwater basins in the kingdom, as follows:

- Yarmouk Basin
- Northern escarpment to the Jordan Valley
- Jordan Valley floor
- Zarqa River Basin
- Central escarpment to the Dead Sea
- West Bank (not part of Jordan since 1988, so it will not be accounted for here)
- Escarpment to Wadi Araba
- Red Sea Basin
- Jafr Basin
- Azraq Basin
- Sirhan Basin
- Wadi Hamad Basin
- Disi-Mudawwara

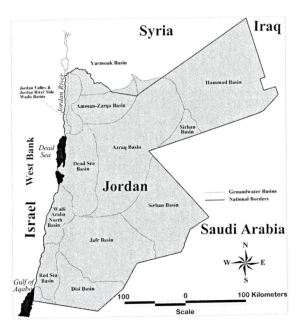

FIGURE 1-3. Groundwater Mean Annual Balance
Source: MWI 2005a.

A detailed description of these groundwater basins is found in Salameh (1996).

Summary of Aquifers and Their Sustainable Yields

Table 1-2 lists the kingdom's renewable groundwater reservoirs. Four of them receive flows from Syria, and as such, they are transboundary reservoirs. Non-renewable (fossil) reservoirs extend under the territories of adjacent countries, particularly Saudi Arabia. Natural recharge of groundwater aquifers was calculated to average 395 mcm/y, and inflow into them from Syria was estimated at 68 mcm/y and is declining because of Syrian upstream abstractions. The total recharge averages 463 mcm/y, and the natural spring outflow is estimated at 197 mcm/y in the form of freshwater base flow. Therefore, the sustainable yield of wells in all aquifers of the kingdom stands at 266 mcm/y. To this should be added the "artificial" recharge coming from the leakage from municipal water networks, return flows from irrigation applied to agricultural lands lying in the catchment of the groundwater reservoirs, and the percolation of wastewater, be it from treatment plants or from cesspools and percolation pits.

The abstraction from the aquifers has been exceeding their sustainable yields for about three decades, since the 1970s. It is evident from the water record that the overabstraction from groundwater wells has permanently affected both the renewable and nonrenewable water resources. From records of the year 2004

TABLE 1-2. Groundwater Reservoirs and Their Estimated Safe Yield (mcm/y)

Reservoir	Inflow from Syria	Recharge	Outflow as base flow	Wells' safe yield
Yarmouk	15	28	10	33
Amman-Zarqa	30	72	40	62
Jordan Valley side wadis	0	110	79	31
Jordan Valley	0	20	0	20
Dead Sea	0	100	56	44
Azraq	18	16	0	34
Hamad	5	10	0	15
Wadi Araba North	0	18	11.6	6.4
Wadi Araba South	0	5	0.1	4.9
Wadi Sirhan	0	5	0	5
Jafr	0	8	0.3	7.7
Disi	0	3	0	3
Total	68	395	197	266

Source: MWI 2004a.

(MWI 2005b), the total groundwater overabstraction from renewable aquifers was determined to be about 173 mcm.[3] Calculations and analyses associated with the writing of this book indicate that this overabstraction has adversely impacted the aquifer water levels, discharge of springs and the resulting base flow, and the water quality extracted from these aquifers (Dweiri et al. 2005). The subject analyses encountered difficulties in discrepancies in the raw field data and insufficient data on base flow and spring flow. However, estimates from the Water Information System (1968–2004) (MWI 2005a) indicate that the average maximum decline in groundwater levels ranged from around 2.53 m in the southern part of the side wadis aquifer to more than 100 m in the Jafr groundwater basin. The Amman-Zarqa groundwater basin sustained level declines ranging from around 40 m in lower elevations to more that 80 m in the higher elevations near Amman and Zarqa. In some parts of the Disi nonrenewable aquifer, the decline in elevation averaged about 20 m as a result of an annual abstraction rate of 62.5 mcm/y. The analyses show that, over the past decade, the overabstraction has also severely affected the spring flow and base flow in side wadis of the Jordan Rift Valley. Erratic data did not allow a reasonable determination of the quantitative impact, but this is being addressed and will be settled before long. The discharge of Azraq springs feeding the Azraq Oasis has dropped by more than 70% (MWI 2004b). Moreover, the Wadi Arab base flow, measured upstream of the Wadi Arab Dam, has all but dried up because of groundwater abstraction from four wells drilled in 1982 to supply Irbid with municipal water. The base flow declined by more than 90%. Elsewhere in the country, and prior to the construction of the Wala recharge dam, only around 25% of Wadi Wala base flow remained after tube wells were drilled to supply Amman and Madaba with municipal water.

A corollary impact of groundwater overabstraction has been the water quality deterioration in practically all groundwater basins by varying degrees (WAJ 2005). In the Amman-Zarqa Basin, for example, the average increase in the concentration of total dissolved solids of some monitored wells was around 700 mg/L, rising to an alarming increase of more that 5,000 mg/L in other locations downstream. In the Hamad and Jafr Basins, the average increase in concentration of total dissolved solids was around 400 mg/L.

In summary, the overabstraction of groundwater in Jordan has crossed red lines by mining aquifers, using nonrenewable aquifers, declining the groundwater levels, reducing the surface base flow, and notably deteriorating the groundwater quality.

Farmers abstracting illegally or in excess of the quantities allotted to them in their abstraction licenses cause appreciable harm to public property (which includes water), the public good, and other users. Harm to public good is manifested by the degradation of water quality of the reservoir, and this has been manifested in the areas of Mafraq, Jafr, Dhuleil, and Amman–Zarqa. Harm to others is displayed in the reduction in base flows as a result of overabstraction. Cases of this type have been manifested in the Zarqa River gorge, Azraq, Wadi Arab, Wadi Jurum, Wadi Wala-Hidan, and Wadi Mujib. Pumping from groundwater reservoirs resulted in the drying up of springs. The Mujib used to have a hefty base flow running under the bridge on the way to Karak from Madaba, but this has dried up because of pumping from Swaqa Reservoir, which supplies Amman with municipal water. In many such cases, public authorities share the responsibility with private well owners for drying up springs, as the case has been in the Amman-Zarqa aquifer, Wadi Hidan springs, Wadi Arab base flow, and Wadi Jurum base flow reductions. However, government authorities mitigated some of the adverse impact by building storage dams to compensate for lost base flows, such as the Wadi Arab Dam raising, King Talal Dam raising, and Wala Dam.

Care was taken to control the runaway groundwater overabstraction. The groundwater management policy of 1998 stipulates that the regulations be strictly enforced. Difficulties arise in law enforcement, especially when the violations support perennial crops and allow needy families to make a living. The turnover in ministers responsible for the water and irrigation portfolio is a major weakness in following up on implementation of the policy and associated rules.

Shared Groundwater Resources

The groundwater resources shared among Jordan and its neighboring countries are found in the fossil reservoir of the Disi aquifer, which extends southward into Saudi Arabia, and more importantly in the renewable aquifers that extend northward into Syrian territories. Along the eastern borders with Saudi Arabia and Iraq, the yields of renewable aquifers are relatively small and generally brackish, receiving only very limited recharge. They are of no significance to either Jordan or its neighbors. The fossil aquifer along the eastern borders had not been explored prior to 2004, when one deep well was drilled; the government policy is to drill more deep wells in that region to investigate the properties of the fossil reservoir and the potential for using it as a source of municipal water supply to Jordan's urban areas. Utilization by Saudi Arabia of this eastern reservoir, as the case has been with the Disi reservoir, could interfere with uses that Jordan may plan in the future for the exploitation of this aquifer.

Along the western borders of the country, in the Jordan Valley area and Wadi Araba, the probability of interconnection between aquifers on both sides of the borders is negligible. The peace treaty between Jordan and Israel regulates the transfer of groundwater from Jordan to Israel in Wadi Araba via agreed-upon conveyance systems in return for Jordan receiving 10 mcm/y of desalinated water in the north Jordan Valley from Israel. Until a desalination plant is erected, this amount has been transferred to Jordan since July 1995 from the Tiberias reservoir in Israel.

In southern Jordan, the fossil groundwater in the Disi aquifer flows from the Aqaba Mountains toward the northeast. The flow lines run almost parallel to the borders between Jordan and Saudi Arabia and gradually turn to the north in central Jordan , and they run parallel to the eastern Jordanian borders with Saudi Arabia and then turn to the northwest. As long as the Saudi extraction sites are far away from the Jordanian borders, no negative impacts on the fossil aquifer underlying Jordanian territories are expected, at least not in the next few decades. Saudi exploitation of this reservoir has been extensive, and the water table around the area of use inside Saudi Arabia has been lowered significantly. In addition, the potential expansion of Saudi well fields northward toward Jordan will bring Saudi projects into areas that had been part of Jordan before 1965 but were exchanged for a Saudi strip of land on the Gulf of Aqaba. The exchange agreement between the two countries in that year states that all the natural resources of the exchanged land are joint property of the two countries. Hence neither Saudi Arabia nor Jordan can unilaterally exploit the water resources in the exchanged territories without the explicit consent of the other party.

In the northern aquifers, the renewable groundwater flows from Syria into Jordan. The Syrian abstractions undoubtedly influence the transboundary flow, decreasing it with time. The groundwater flow from Syria into Jordan across the borders, according to some sources, was estimated at 20 to 30 mcm/y (Salem 1984). The Ministry of Water and Irrigation updated those estimates and found them to be about 68 mcm/y, as shown in Table 1-2. Exploitation of groundwater on the Syrian side is expanding and is expected to reduce the transboundary flow into Jordan by about 50%.

TABLE 1-3. Groundwater Mean Annual Balance (mcm/y)

Budget component	Annual quantity
Groundwater recharge from precipitation	395
Transboundary groundwater flow from Syria	68
Network leakage, cesspools, and irrigation return flow	70
Total inflow	533
Groundwater abstraction (wells)[a]	440
Base flow in various watercourses	197
Total outflow	637
Change in storage (inflow-outflow)	(−104)

a. A total of 85 mcm was abstracted from fossil reservoirs at Disi and is not included in the total groundwater abstraction. When this is added to the overabstraction, the change in storage becomes 189.
Source: MWI 2004a.

The mean groundwater budget shows that the renewable reservoirs were used over and above their total sustainable yields by 104 mcm. The overabstraction in 2002 was 179 mcm, 75 mcm more than the mean above, indicating that the overabstraction of groundwater from wells has not been arrested despite all the intentions of the groundwater management policy (1998) and the legislations of 1977, 2002, and 2004 penalizing the overabstraction through higher water tariffs.

Precipitation Water Balance

Table 1-3 summarizes the groundwater balance. The inflow into the groundwater reservoirs consists of natural recharge from precipitation over the kingdom's territories at 395 mcm/y, a transboundary component from precipitation over southern Syria of 68 mcm/y, and return flows from irrigation and municipal and industrial uses of 70 mcm/y, all totaling an average of 533 mcm/y. The total annual outflow from them in 2002 was about 637 mcm in the form of base flow (197 mcm) and well extraction (440 mcm). The water balance of groundwater is in deficit every year at an average of 104 mcm, causing reductions in base flow of side wadis and lowering groundwater tables.

The indigenous component of groundwater safe yield excludes the exogenous (68 mcm from Syria) and the return flows (70 mcm) and is therefore only 198 mcm/y. The total indigenous blue water consists of the surface-water component (554 mcm/y) and the sustainable groundwater component (198 mcm/y) and adds up to 752 mcm/y.

Green water is stored in the soil by capillary action and surface tension. Only plant roots can make beneficial use of this water, and evaporation takes care of it if not so used. The blue-water equivalent of green water in Jordan was calculated to average about 866 mcm/y (Chapter 7).

The precipitation balance for Jordan is expressed by the simple linear equation

$$P = B_w + G_w + Ev,$$

where

P is the total average annual precipitation over the country (8,200 mcm/y),
B_w is the blue water (752 mcm/y),
G_w is the green water (866 mcm/y), and
Ev is the evaporation, in mcm/y.

Solving the above equation for the unknown evaporation,

$$Ev = P - (B_w + G_w)$$
$$= 6,582 \text{ mcm/y.}$$

Thus the rate of annual evaporation, Ev/P, amounts to 80.28%, a high rate indeed, which places Jordan in the arid and semi-arid zone.

The world total average precipitation is about 111,000 km³, of which 39,000 km³ become blue water, 8,800 km³ become green water, and the balance of 63,200 km³, or 56.9%, is evaporated (WRI 1988). Jordan's evaporation rate is about 141% of the world average.

Summary of Water Resources Potential

Additional water resources have been used in Jordan in 2004 over and above the renewable indigenous resources of 1,618 mcm, which is the sum of indigenous blue water (752 mcm) and green water (866 mcm). The additional resources consist of the transboundary groundwater originating in Syria (68 mcm); surface water originating from neighboring countries at 284 mcm/y (Table 1-1); 80 mcm of gray water (treated wastewater from treatment plants); and 70 mcm return flow from water network leakages, irrigation return flow to aquifers, and wastewater infiltration from local treatment facilities (Table 1-3). Nonrenewable uses are discounted. Thus the total available renewable water resources consist of

$$W_R = B_w + G_w + T_w + L_w + I_w,$$

where

W_R is the total renewable water resources potential,
B_w is sustainable indigenous blue water (a total of 752 mcm),
G_w is the green water (866 mcm),
T_w is the treated wastewater from treatment plants (80 mcm),
L_w is the leakage water from water supply systems and local wastewater treatment facilities (70 mcm), and
I_w is the exogenous surface water (284) and groundwater (68), totaling 352 mcm/y.

The above equation yields total water resources potential of 2,120 mcm/y. This total is sensitive to the exploitation of green water, which was calculated assuming cultivation of 3.13 million dunums (du) under dry farming conditions and 7.91 million du of rangeland (Chapter 7), and to the rainfall quantity and frequency.[4] It is also affected by the inflow into groundwater reservoirs from domestic water supply networks, and this quantity is being reduced with time. Whereas leakage of locally treated wastewater is expected to diminish with expanding coverage of wastewater collection networks, the surface flow from wastewater treatment plants will increase.

The nonsustainable uses totaled 189 mcm in the form of overabstraction from groundwater (104 mcm) and the use of fossil water (85 mcm). Although overabstraction is destined to diminish, the use of fossil water is expected to increase in the future.

The population–resources equation is best displayed by the per capita share of sustainable water resources as compared to the need for water resources per capita. The water need per capita, or water threshold, is 1,700 m³ per capita for a country of the lower-middle income category like Jordan, consisting of 1,500 m³ for agricultural needs, 75 m³ for municipal needs, and 125 m³ for industrial needs (see Table 7-4).

The above summary of water resources indicates that in 2004, the sustainable water resources were 2,120 mcm, and the population was 5.35 million. The per capita share of sustainable water resources was therefore 396 m³, as compared to a need of 1,700. Thus in 2004, Jordan had only 23.3% of its water needs, a ratio that reveals how imbalanced the population–water resources equation for the country has been.

Future Water Demand

The overall water quantity used in 2002, including soil water, was 250 mcm for municipal uses, 37 mcm for industrial, and 1,377 mcm for agricultural. The total was 1,664 mcm, or 312 m³/cap, compared to the total need for all three purposes, calculated at 1,700 m³/cap/y (Haddadin 2003) (see also Table 7-4). This use, including overdraft from groundwater reservoirs and the reuse of treated wastewater, amounted to 18% of the water needed for municipal, industrial, and agricultural purposes. The use constituted 73.5% of the renewable water potential. Future water demand and allocation among domestic, industrial, and agricultural uses in the country also can be projected.

Domestic Uses

The supply of municipal water in the kingdom in 2002 was 250 mcm. Accounting for population increases at the rate of 2.8% per year, the supply would have to be 310 mcm in 2010 and 410 mcm in 2020. If the supply rate is improved to 70 m^3 per capita, the amounts needed would be 435 m^3 in 2010 and 615 m^3 in 2020. The total annual flow of fresh blue water, including the international shares, was shown to be 1,066 mcm, composed of 800 mcm of surface water (the portion developable from the 835 mcm total surface-water resources) and 266 mcm of groundwater. The municipal demand, if fully met from indigenous sources, will, in 2020, claim 57.7% of the freshwater resources of the country. This will escalate the competition among the different users and could create tension with the neighboring countries if cooperation among them is not assured to preserve the quality of shared water and honor the rightful shares.

Industrial Uses

The 2002 level of industrial uses was 37 mcm (7 m^3/cap). The forecast is about 55 mcm for 2010 and 80 mcm for 2020. Industrial uses will thus claim 7.5% of the freshwater resources in the year 2020. The main increase is expected to be caused by the Dead Sea chemical industries and oil shale extraction and processing.

However, the industrial water needs are substantially more than the actual amounts used by Jordanian industries. As is the case with agricultural water, the deficits in industrial water needs are met by imports. It is realized that Jordan cannot produce all the industrial commodities it needs because of the size of the domestic market, lack of technological capacity, limited capital, and research constraints. But assuming a virtual environment under which domestic production of all manufactured needs is possible, the average industrial needs for Jordan as a lower-middle income economy is calculated to be around 125 m^3/cap/y (Table 7-4). The balance between this need and the amount used (7 m^3/cap), or 118 m^3/cap, was provided through industrial imports.

The combined demand for municipal and industrial water in 2020 amounts to 695 mcm, or 65.2% of the freshwater resources. When priority of allocation is given to those uses, agriculture would be left with little freshwater resources and would have to rely more on treated wastewater for irrigation and green water (dry-land farming) for food production. The population distribution over the territories of the kingdom will be adversely affected when agricultural areas shrink because of irrigation-water diversions to other uses. People will follow the water, and migration to urban areas from the agricultural rural areas will ensue.

Agricultural Uses

The amount of water used in irrigation depends on the availability of resources. In 2002, this sector consumed about 1,377 mcm, consisting of 693 mcm (80% of 866 mcm) of green water, 523 mcm of blue water from surface and groundwater resources (including about 167 mcm of overabstraction), 72 mcm of gray water (treated wastewater), and 89 mcm of fossil water. The total average was 258 m^3/cap. The agricultural need was 1,500 m^3/cap (Table 7-4). The deficit, or 1,242 m^3/cap, was covered by food imports. The agricultural water deficit would have been 1,290 m^3/cap had only the sustainable flows been used that year.

Agricultural water will be augmented with treated wastewater reuse and desalination of brackish water. Private farmers have embarked upon this in the Jordan Valley. A total of 23 private small desalination plants have been installed and are operational, with an aggregate capacity of 1,000 m^3/ hr.

Future Prospects

It is clear that the population pressure on water resources in Jordan is high and continues to rise. Possibilities for augmenting water resources are limited. The Wehda Dam on the Yarmouk River will increase surface-water availability in the coming years, as will greater treated wastewater flows, a function of expansions of municipal water supplies and wastewater collection networks.

The following are expected to augment the supply of municipal and industrial water:

- desalination of brackish groundwater and surface water (four desalination stations with an aggregate capacity of 5,500 m^3/hr are in operation, owned and managed by the Water Authority of Jordan);
- desalination of surface-water resources (Mujib, Zara, and Zarqa Ma'een, now being implemented);
- surface-water sources from Wehda Dam on the Yarmouk; and
- fossil-water sources (Disi aquifer).

Another source of municipal water could be seawater desalination, but only when the national income is improved to the extent where water consumers could afford its cost without treasury subsidies. The current stated policy is for seawater desalination to augment the municipal supplies for the port city of Aqaba in the future.

There has clearly been an increase in intensity of competition over water resources, especially those of drinkable quality with some degree of treatment. The likely victim of any reallocation of water resources will be agriculture, to the advantage of municipal supplies.

Summary

Jordan's population grew 11.5 times in 66 years, from 0.3 million in 1938 to 5.35

million in 2004, because of the abrupt influx of population in the wake of the turbulence that has been affecting the Middle East. The renewable water resources of the country originate from precipitation over its territories (8.2 bcm), of which only 20% can be used in beneficial purposes, and the rest is lost to evaporation. Only 3.1% of Jordan's territories receive more than 300 mm of rainfall.

The precipitation over Jordan flows into seven distinct drainage basins, most important of which is the Dead Sea Basin, with renewable surface-water resources averaging 554 mcm/y; an average of 284 mcm/y of surface flows accrue to the country from neighboring Syria (184 mcm) and Israel (100 mcm). The infiltration into the groundwater tables takes place in 12 renewable groundwater reservoirs existing in the country. The sustainable yield of all the aquifers amounts to 463 mcm/y, of which 68 mcm are subsurface inflows from Syria, 197 mcm emerge as spring flows (counted in the surface resources above), and 198 mcm are aquifer recharge from precipitation on Jordanian territories (indigenous recharge). The total indigenous blue-water resources amount to 752 mcm (554 + 198). Thus the total blue-water resources amount to 1,104 mcm/y, of which an average of 352 mcm, or about 32%, come from the neighboring countries. When green-water potential is added (866 mcm), the total freshwater potential of Jordan amounts to 1,970 mcm/y.

The groundwater component of the blue water has been overexploited at an average rate of 104 mcm/y, with the actual figures rising over time despite government's efforts to arrest the overabstraction. Surface-water exploitation entails expensive dam construction projects. By 2002, about 434 mcm of surface-water resources have been utilized, with five dams built at a live storage capacity of 110 mcm. More dams were being constructed, adding more storage capacity and yield.

Gray water and return flows contribute a total of 150 mcm to the usable water resources, making the grand total of renewable water resources 2,120 mcm. For a population of 5.35 persons (2004), the per capita share was 396 m^3, as compared to a need of 1,700 m^3/cap, or 23.3% of the needs.

The future forecast of water allocation calls for the allotment of 65.4% of the renewable freshwater resources to municipal and industrial uses, leaving the rest of the blue water to irrigation. The 2002 usage for all purposes amounted to 73.5% of the total renewable potential and supplied only 18% of the water needs in the country. The remaining 82% of the needs were covered through imports.

Future prospects for increased water supplies for municipal and industrial uses focus on desalination of brackish water from ground and surface resources, impounding the flood of the Yarmouk at the Maqarin site (Wehda Dam), and exploiting the strategic reserves of Disi water.

References

Baker, M., Jr., Inc., and Harza Engineering Company. 1955. *Yarmouk Jordan Valley Project: Master Plan Report.* Amman, Jordan: Ministry of Finance, submitted to Jordan–United States Technical Services.

DOS (Department of Statistics). 2004. *Census 2004.* Amman, Jordan: DOS.

Dweiri, S., H. Naber, and M. Haddadin. 2005. Impacts of Over-abstraction on the Quantitative and Qualitative Behavior of Groundwater Aquifers in Jordan. Paper in preparation for submission to *Water International,* 2006. Draft available from the editor of this volume.

GTZ (Gesellschaft für Technische Zusammenarbeit). 1977. *National Water Master Plan.* Amman, Jordan. Prepared for the Natural Resources Authority.

Haddadin, M. 2001. *Diplomacy on the Jordan: International Conflict and Negotiated Resolution.* Norwell, MA: Kluwer Academic Publishers.

———. 2003. Significance of Shadow Water to Irrigation Management. Paper sponsored by the University of Oklahoma, Norman, and Oregon State University, Corvallis, submitted to the Prince Sultan Bin Abdulaziz International Prize for Water.

MWI (Ministry of Water and Irrigation). 2004a. *Mean Annual Water Budget.* Amman, Jordan: MWI.

———. 2004b. *National Water Master Plan (from 1963 to 1999),* vol. 3. Amman, Jordan: MWI.

———. 2005a. *Water Management Information System.* Amman, Jordan: MWI.

———. 2005b. *2004 Water Balance.* Amman, Jordan: MWI.

Qasim, S. 1993. *Arab Food Security: Present and Future.* Amman, Jordan: Shoman Foundation.

ROID (Regional Office for Integrated Development). 1993. *Jordan River Basin Study,* part 1, vol. 1, *Main Report.* Amman, Jordan. Submitted to the World Bank, 36–41.

Salameh, E. 1996. *Degradation of Water Quality in Jordan.* Amman, Jordan: Royal Society for the Conservation of Nature in cooperation with Friedrich Ebert Foundation.

Salem, H. 1984. Hydrology and Hydrogeology of the Area North of Zarqa River. M.Sc. thesis, University of Jordan, Amman, Jordan.

WAJ (Water Authority of Jordan). 2005. WAJ laboratory results database from 1960 to 2005.

WRI (World Resources Institute). 1988. *World Resource (1988–1989).* Washington, DC: WRI and UNEP, Tables 17.2 and 18.5.

Notes

1. Initial results of the 2004 census revealed that the population level reached 5.35 million that year, compared with the forecast level of 5.6 million. Reference in this chapter is made to the forecast levels of population.

2. Figures quoted from the works of Ambassador Eric Johnston's mission to the region

(1953–1955), a copy of which was made available to the author in 1984 by Dr. Wayne Criddle, a member and engineer of the mission.

3. In 2004, 62.5 mcm was abstracted from the nonrenewable Disi Aquifer.

4. One dunum is one-tenth of a hectare or 1,000 m².

2

Evolution of Water Administration and Legislation

Munther J. Haddadin

Water policy is not disconnected from the various other policies that a country may adopt from time to time. It is part of the overall state governance and should blend with other policies as the case may dictate. Especially when water resources are shared with other countries, water policy closely interacts with diplomacy, foreign policy, and policies on cooperation with the co-riparian party in other sectors of the economy. No wonder, then, that water policy gets overhauled, amended, or its emphasis shifted in light of regional political and economic interactions and developments. On the domestic front, shifts in water policy usually take place in light of the economic and social development potentials and the need to achieve higher rates of economic growth and progress. Income distribution patterns, and the responsibility of the State in that regard, may also be a key factor in shifting water policy toward a more balanced pattern of allocation and re-allocation. Water, in simple terms, is an indispensable input for the generation of wealth, and its allocation to consumers and purposes of use affect income distribution patterns. Impacts of water policy touch on all aspects of economic and social activities and environmental well-being.

Legislation is the tool for implementation of water policy in Jordan. When water legislation is enacted, it reflects the water policy adopted at the time, and any changes in that are enforced through amendments of the laws in effect. Water legislation assigns the responsibility of implementation to an existing entity or a new entity that the same law establishes. Thus the institutional arrangement is spelled out in the legislation that, in turn, is made to implement a policy.

The hierarchy of legislation in the country has as its peak the Constitution, which is adopted through votes of the elected lower house and appointed upper house of Parliament. Next in the hierarchy of legislation come the laws, conforming to the relevant provisions in the Constitution, approved by the Council of Ministers, debated and passed by both houses of Parliament, and issued for enforcement by a royal decree. Detailing of the intent of laws is done through regulations or bylaws whose provisions conform to those of the respective law. The bylaws need to be approved by the Council of Ministers only, as they conform to the law that the representatives of the people (both houses of Parliament) had approved earlier. A royal decree is issued to make the bylaws go into effect. Finally, the administrative decisions made by the authorized administration official form the last link in the legislative chain. Those decisions should conform with the provisions of the bylaws, the laws, and the Constitution, as the case may be. The Supreme Court has the constitutional authority to look into disputes resulting from nonconformity of administrators of any entity with the legislation in force at the time of the administrative decision.

In this chapter, legislation and administration as required for implementation of water policy are reviewed, and the links between the water policy and the overall strategy of the country are examined.

The Genesis of Water Policy in Jordan

The Emirate of the Arab East (1921–1923) and its successor, the Emirate of Transjordan (1923–1946), gradually built a state infrastructure with an Executive Council and later developed an elected Legislative Council. The judiciary branch was independent and run by the courts. The emirate received financial and technical aid from Britain and was assisted by British civil and military administrators. Legislation in the emirate consisted of the ongoing laws that organized life under the Ottoman Empire, as well as new laws borrowed from next-door Palestine, then under the administration of the British Mandate.

Water administration was focused on municipal water supply, a responsibility assigned inside towns to the municipal councils. The role of central government was minimal, mainly to facilitate implementation of the projects, legislate for the service, and guarantee loans extended to implement the projects. Elsewhere in the emirate, where there were no councils, villagers depended on gravity wells in which rainwater was collected or on springs for their municipal water sources.

Water policy expanded to engulf groundwater and irrigation water in the wake of the Palestinian revolt of 1936–1937 against Jewish immigration into Palestine. Groundwater was a target for the Jewish Agency, and so were the surface waters of the Jordan Basin. Two laws were enacted in Palestine to reduce frictions between Jews and Arabs, emphasizing the authority of government over water resource controls and investigation. The emirate borrowed the two laws and applied them.

British royal commissions came to the region to find a resolution to the Palestinian–Jewish conflict and pacify the Arab anger. They recommended partitioning Palestine between the Palestinian Arabs and the Jews and having the Palestinian portion amalgamated in the Emirate of Transjordan (Ionides 1946). The recommendation prompted the government of Jordan to look into the sufficiency of water resources for such an eventuality.

Michael Ionides, the British manager of the Projects Department in Jordan and an experienced water engineer, conducted a 1939 study to assess the adequacy of water resources of the Jordan Basin to support a two-state solution (Ionides 1946). The study marked the first involvement of the emirate in what could then be termed as international water issues. Jordan's policy was clear: to defend the water rights of the indigenous populations of Palestine and Transjordan. The Ionides study helped form the basis of Jordanian and Arab policies toward the waters of the Jordan River Basin. In 1938, Ionides had prepared a draft law known as the Water Supervision Law to organize supervision of water resources in the emirate, but his departure to join the troops in World War II halted its enactment until 1952.

It is clear that Jordan's water policy and its translation into laws came as a reaction to plans for the establishment of a "national home" for the Jews in Palestine, as pledged by the British government in 1917, culminating in the establishment of Israel in 1948.

Water Administration and Legislation in the Kingdom

Jordan earned independence in 1946 and was proclaimed a kingdom. Its Basic Law stipulated three branches of government: executive, legislative, and judiciary. The West Bank, the portion of Palestine that was defended and retained by the Jordanian and Iraqi troops during the 1948 war with Israel, united with the kingdom in 1950. A basic overall policy of the kingdom had been coordination with its Arab sister countries, bilaterally or multilaterally through the League of Arab States, in all matters relating to or affecting overall Arab interests, especially those touching on Palestinian rights and matters related to Israel.

Plans for the Yarmouk

The water policy of the kingdom at the outset focused on the Jordan River Basin in response to a similar focus accorded to the basin by the newly founded state of Israel. In 1949, the government commissioned a British consulting firm, Sir Murdoch MacDonald and Partners, to appraise potential use of the waters of the Jordan River Basin. Its report, favorable to the kingdom, was published two years later (MacDonald and Partners 1951). The Ionides study along with the MacDonald report formed the takeoff platform for the master plan that the government would later formulate for the development of the Jordan Valley (Baker and Harza 1955).

Between 1946 and 1957, the Projects Department within the Department of Lands and Surveys, under the minister of finance, was in charge of water resources. The regulation and control of these resources were to be exercised by the director of lands and surveys, aided by the manager of the Projects Department under him. The water rights for irrigation went hand in hand with the ownership rights of the land to be irrigated. The Islamic Sharia was the source of legislation as far as water rights were concerned, and the rules issued by the Majallah, the Ottoman Official Gazette, were the major guide in that regard. Ottoman legislation, basically in conformity with Islamic Sharia, had been the law of the land for four centuries, and its momentum carried through the kingdom era. As was the case during the mandate period, the local councils of towns and cities remained in charge of the domestic water supplies.

The influx of Palestinian refugees into Jordan created pressures for jobs and means of livelihood. Jordan's economy was weak, and foreign assistance was essential to boost development and create jobs; water was critical for the development process. Because of the urgency, a department for the development of water resources was established and staffed with American experts and Jordanian employees (Dokhgan 2004). Among such water experts was the American engineer Mills Bunger, whose ideas were crucial to the formulation of Jordan's policy concerning the Yarmouk River and the Jordan Valley development. Bunger recommended the regulation of the Yarmouk flow by a dam on the river course at Maqarin, unlike all other previous plans, which identified Lake Tiberias as the likely regulation reservoir. The Jordanian government endorsed Bunger's idea, which tied in with its policy of protecting its water rights, preferring a reservoir on Arab lands like Maqarin to a reservoir in enemy hands like Lake Tiberias. The Yarmouk has Syria upstream for a riparian party and Israel downstream for another.

Without consulting with either of the two riparian parties, Jordan, with financial support from the United Nations Relief and Works Agency for Palestine Refugees (UNRWA) and the U.S. Technical Cooperation Agency (TCA), started site investigation works for the Maqarin (Bunger) Dam in 1952. In response to Israeli complaints to the United States over its being ignored as a riparian party, the financial support was withdrawn (Haddadin 2001), but the project remained the central piece in the Jordanian and Arab position with regard to the development of the Yarmouk River.

Jordan's eagerness to have the dam built on the Yarmouk was justified by its need to irrigate as much as it could of the Jordan Valley and create jobs for an enlarged agricultural population. In pursuit of cooperation and riparian consultation, Jordan approached Syria and concluded a bilateral agreement in June

1953 to utilize the Yarmouk River. Cooperation or direct consultation with Israel was unthinkable then, as the Arabs and Israel were declared enemies.

The interest of UNRWA in supporting the Yarmouk project was clear. It was in line with its mandate to provide work for the Palestinian refugees. The interests of the United States, the leader of the free world during the cold war, were, in brief, to combat poverty and want, as those problems would encourage the infiltration of communism into the sensitive and strategic region of the Middle East, and to settle Palestinian refugees in the Jordan Valley as part of a solution to the Arab–Israeli conflict.

The competing water plans of Jordan and Israel eventually led to conflicts and military clashes, especially between Syria and Israel. The heated environment prompted U.S. president Dwight Eisenhower to dispatch Ambassador Eric Johnston to the region to work out a unified plan for the development of the Jordan Valley to which both the Arab states and Israel could subscribe. The plan, finalized in 1955, became the basis of Jordan's policy with regard to the waters of the Jordan Basin. (Details are given in Chapter 10.)

Legislation for Water Rights

In the country were water springs, perennial streams, ponds, and river flows whose ownership status and mode of usage were not properly defined or officially documented. The government decided to update the landownership records and adopt modern surveying techniques to replace the archaic methods of the Ottomans. The Palestine Grid, established by the British Mandate in Palestine, was extended to cover the emirate (later, kingdom) with triangulation points over its territories as necessary. The origin of coordinates of the Palestine Grid was at Gaza, the southwestern city in Palestine.

Lands already irrigated with water from springs, streams, or rivers had to be so designated in new title deeds that the Department of Lands and Surveys would issue in replacement of the deeds of Ottoman time. The amount of water needed to irrigate that land had to be specified as well. For that purpose, the first water law enacted in the kingdom responded to the policy set to regulate water use rights. It was law number 38 for the year 1946, the Law of Settlement of Water Rights, by which the procedures for settlement of water rights were specified. The law authorized the prime minister, upon recommendation of the director of the Department of Lands and Surveys, to designate as an "irrigation area" any area in which the water rights had to be settled. The department would either adopt the irrigation system that was in place or build a new one, and the law authorized the director to construct, operate, and maintain the irrigation projects in such areas and collect water charges from the beneficiaries. The law obviously entrusted the government to manage water resources, a measure that was viewed as crucial to prevent local disputes among farmers over water rights.

Because irrigation was the primary user of water resources, the settlement of water rights was linked to the irrigable land, and a further law combining both land and its irrigation water was issued: law number 40 for the year 1952, the Law of Settlement of Land and Water Rights. In addition to this major task of settlement of rights, the director of the Department of Lands and Surveys, aided by the Projects Department under him, was given the responsibility, through law number 31 for the year 1953, the Water Supervision Law, to construct and manage the irrigation projects, provided that the owners of its lands carried two-thirds of the capital cost. That law had been drafted by Michael Ionides before he departed for military service.

To administer and manage groundwater, a well-drilling department was initiated at the Ministry of Public Works. It was clear that the private sector was not prepared to lead in the introduction of drilling technology, and that the government had to do so. The government was eligible for technical assistance from friendly countries and was eager to sound and assess the groundwater resources. Additionally, because of the connections between groundwater and base flow of wadis that farmers were using, it was important to make sure that future well drilling would not harm existing uses of base flows.

Focus on the Yarmouk and the Jordan Valley

Between October 1953 and October 1955, water was accorded much attention not only in Jordan, but also in the rest of the member countries of the Arab League. This was the period of shuttle-water diplomacy conducted by the American presidential envoy Eric Johnston over the waters of the Jordan River Basin.[1] Jordan took the initiative and made contacts with Syria, with whom it concluded a treaty over the utilization of the Yarmouk River on June 4, 1953. This treaty showed the priority that water had for Jordan and a parallel priority that Syria placed on power generation from the dam. Water priority for Syria manifested itself in the treaty as well. An autonomous Joint Committee for the Utilization of the Yarmouk River was set up between the two countries.

The water projects received central attention, and the Jordanian government decided in 1957 to shift the responsibility for them to the Ministry of Public Works. This was done through law number 22 for the year 1957, which amended law number 31 for the year 1953. The takeover by the Ministry of Public Works proved to be only temporary, and government policy shifted gears again to accord water the priority and attention it deserved. The country was under political pressure from Syria and Egypt, and it very much needed to reduce unemployment to safeguard against radicalization of its young people agitated by hostile broadcasts from the two neighboring countries.

Water Policy Shifts Gears

The year 1950 saw the challenges facing Jordan's marginal economy compounded. The West Bank became an integral part of the kingdom, carrying with it not only a mass of population, but also substantial defense burdens. The armistice line between Jordan and Israel became the longest front of military confrontation. Economic and social development were essential to achieve an acceptable quality of life for the expanded population, and water resources were vital.

Ambassador Johnston's shuttle culminated in an agreement at the technical level on water sharing and the development of the waters of the Jordan River Basin. It paved the way to develop the Jordan Valley in the Hashemite Kingdom of Jordan and the water resources at large. This development became the cornerstone of Jordan's water policy for several reasons: water was required for the urgently needed development; exploration and survey of water resources were essential for proper planning; the Jordan Valley Project was viable on economic, social, and political grounds; and benefits from the Jordan Valley could accrue almost immediately after the project, or any part thereof, was completed. Jordanians (and Palestinians for that matter) needed little or no training to practice irrigated agriculture.

Jordan, encouraged by the prospects of reaching an agreement on water sharing, and with financial support from the United States Agency for International Development (USAID), commissioned a joint venture of two American firms in 1953 to prepare a master plan for the development of the Jordan Valley. The plan was submitted in 1955, but the political turmoil in the region required caution, and government attention was devoted to security issues. In 1958, attention once more turned to the development of water resources and the Jordan Valley. The United States expressed preparedness to support Jordan's efforts in the development of the Jordan Valley, provided that it would not draw more water from the Yarmouk than the share specified in the Johnston Plan (MOFA 1958). No conditions were ever made to ensure the settlement of Palestinian refugees in the Jordan Valley.

This development, along with the urgent need for economic growth and job creation in a turbulent political regional environment,[2] called for a shift in the water policy. It also was fueled by competition over water resources with Israel. Jordan and Israel each had its own plans regarding the Jordan and Yarmouk rivers. Only the Johnston Plan made the two plans reasonably non-contradictory. However, Jordan feared that Israel, which had superior capacities to build, might beat it and transgress on its water shares. The policy shift was visible in the size of investment allotted to the water sector, with generous grants from the United States, and in the unprecedented legislation that was enacted to translate the policy into action on the ground. Furthermore, the institutional arrangements that were set up to implement the policies testify to the drastic shift in policy of water administration and management. The strategic importance of the Jordan Valley dictated that special arrangements be set up for its development. The encouragement of the donor, the United States, helped accelerate the steps to implementation.

Legislation for the New Policy Trends

Legislation was enacted to implement the new trends in water policy. Law number 14 for the year 1959, the East Ghor Canal Law, created an autonomous authority for the first time in Jordan, supervised by a board of directors. The East Ghor Canal Authority (EGCA) was headed by a competent engineer, Nicholi Simansky, a naturalized British citizen originally from Belarus. He was aided by several Jordanian engineers and technicians.[3] The second was a law creating a central water authority, described below.

The East Ghor Canal Law reflected unprecedented policy of irrigated-land reform. It undoubtedly was drafted in collaboration with USAID as a tacit cooperative gesture between the United States and Jordan. The land to be irrigated was to be subdivided into farm units of 3 to 4 hectares (ha), and the law would redistribute the irrigated land to owners and landless farmers. USAID believed that such measures would give the Palestinian refugees a good chance to own irrigated land and thus prompt them to settle, and the Jordan government saw the provisions as an opportunity for land reform, a popular path that both Egypt and Syria (with revolutionary regimes hostile to Jordan at the time) had followed. Despite land reform patterns implemented in Egypt and Syria at the time, the Jordanian policy was more compassionate with landowners.

Under this law, the irrigation project area would be expropriated and the value of the land assessed and recorded as a capital asset in favor of the owner. After the project was implemented, the land would undergo geometric subdivision into farm units of 3 to 4 ha each. An owner with less than 4 ha would retain an area equal in value to his recorded capital assets. If the area was greater, he was allowed to retain a percentage of the portion in excess of 4 ha in descending order as the ownership increased. The more his land area before the project, the less would be the percentage of land he was allotted after the project. Where the land was developed with tree plantations before the project (with perennial flow of side wadis), the owner, with the approval of the Council of Ministers, could retain his entire developed ownership. Lands in excess of the owner's right, in addition to state land in the project area, were allotted to landless farmers. Two amendments, laws number 13 and 31 for the year 1962, were introduced to adjust certain provisions as a result of actual experience of applying this policy on the ground.

In a way, the authority went into partnership with the landowners to develop the land and gave itself the right to dispense with part of the privately owned land (in addition to state land) for the benefit of landless farmers. The authority, in return, carried all the capital cost of development and subsidized the operation and maintenance costs, including water tariff. The government achieved the social gain of the allotment of irrigated land, against payment in installations, to landless poor farmers.

This policy of irrigation development in Jordan proved to be a deserved success. On the one hand, its social and economic implications were extremely positive, and on the other, its political effect on preserving Jordan's rights to Yarmouk waters was documented and exercised. The landowners, although prohibited from selling any part of their ownership to others, including members of their own families, were not unhappy to lose part of their ownership, which had been unproductive in the arid Jordan Valley. The landless farmers who were entitled to landownership were eager to practice farming not as tenants but as owners, a status that gave them social recognition and entitled them to borrow for agricultural development with their holding as collateral. The valley transformation encouraged its population to settle therein, and it even attracted residents of outside districts to live there and assume activities in services supporting an agricultural community. (Details of social gains are presented in Chapter 9.)

Centralizing Water Administration

A second law enacted in the course of water policy shifting gears was concerned with the entire kingdom, not just a region of it like the East Ghor Canal Law. Law number 51 for the year 1959 was meant to reorganize the management of water resources in the kingdom, creating the Central Water Authority (CWA). This new autonomous government agency consolidated the multiple departments dealing with water: the Department of Irrigation; the Drilling Department at the Ministry of Public Works; the Department of Water Resources Development at the Office of Consolidated Services, which merged with the Jordan Development Board; and all operational water projects handled by any other department. The CWA was responsible for all water matters in the kingdom except the East Ghor Canal Project in the Jordan Valley and the municipal water supply distribution networks.[4]

This policy and the associated law proved to be very beneficial to the country, enabling a professional approach to the water sector with particular emphasis on resource surveys and exploration. Benefits included the documentation of resources, measurement of flows and quality parameters, drilling for groundwater investigation, and municipal supplies to needy cities. This measure was a first in the region. Neighboring countries had ministries in charge of water, mostly more than one ministry handling the water affairs, whereas Jordan's agency was autonomous and allowed for better efficiency. Both the East Ghor Canal Authority and the Central Water Authority operated apart from government routine and offered better pay, so they attracted competent staff and inspired dedication. Each authority had its own procurement regulations, with powers given to their executives to handle urgent matters and emergencies. Politicians became critical of the setup, however, as the authorities made reasonable progress without political controls.

Administrative and Legislative Cooperation with the Arab League

Water policy would make yet another major shift, this time as a result of the management of international waters shared by Arab riparian parties on the Jordan River Basin. The shift in policy was dictated by strategic considerations. An Arab Summit meeting, the first ever since the loss of most of Palestine to Israel, was convened in Cairo in January 1964 to consider a unified Arab position toward Israel's unilateral act of diverting substantial flows of the Jordan River to the Negev in Israel's dry south.[5] A second summit was convened in Alexandria, Egypt, in September of that year, and a plan was adopted to divert flows from the upper tributaries of the Jordan River, originating in Arab lands, and leave the rest for Israel's use. The Arab move was a late reaction to Israel's National Water Carrier project to transport Jordan Basin water to areas inside Israel outside the basin. The shift in Jordan's water policy was needed to coordinate efforts with the Arab riparian parties and the League of Arab States.

The Arab League introduced a regional organization and entrusted it with the implementation of the Arab Diversion Project. For this purpose, a new agency was established in Jordan, the Jordan River Tributaries Regional Corporation, created by law number 11 for the year 1965. This autonomous agency was part of the above regional organization decreed by the Arab League and operated in Syria and Lebanon as well. In Jordan, it reported to the prime minister and was staffed with Jordanian officials. The work of this agency, and its sister agencies in Lebanon and Syria, was aborted by the June 1967 war in which Israel occupied the Syrian Golan, where the major components of the Arab Diversion Project were located.[6] With the continued stagnation in its work after the war, the corporation's focus shifted to the domestic front, and it became in charge of planning and implementing dams on Jordanian watercourses in the Jordan River Basin. It finally was incorporated in 1977 in the Jordan Valley Authority as the Department of Dams.

Consolidation of Water Agencies

The need arose to consolidate the departments working in the field of natural resources, including water, into one government agency. Law number 37 for the year 1966 created the Natural Resources Authority (NRA) and defined its responsibilities and authorities. Under its jurisdiction was a merger of the East Ghor Canal Authority, the Central Water Authority, and the Department of Mining, which had been created in the Ministry of National Economy. The NRA was further empowered to build distribution networks in all the population centers in the kingdom, with the exception of Amman. Because of the importance of natural resources development, the new authority had a board of directors chaired by the prime minister as president and was actually run by a vice president who enjoyed a high rank in the government.

Shortly after the NRA was established, the June 1967 war broke out, and the agency's activities were mostly confined to the operation and maintenance of the projects that were under its constituent predecessors. One development project that the NRA pursued was the extension of the East Ghor Canal by 8 km, from station 70 to station 78. A general stagnation of development activities occurred in the region because of the lack of internal stability in the wake of Israel's occupation of Arab lands, including the West Bank, and the emergence of the militant factions of the Palestine Liberation Organization (PLO), which operated against Israel from the Jordan Valley. Retaliatory attacks by Israel all but depopulated the valley and almost brought to a halt the agricultural activities there.

Policy Shifts after the June 1967 War

The government of Jordan, upon direction by the king, turned its attention to economic and social development in early 1971 after law and order were reestablished. It was mandatory to resume economic and social development, which had remained dormant for almost four years after the war of 1967. Efforts were needed to make up for lost time and prepare development projects. A change in water policy ensued, which touched on the philosophy of water development and uses. It added municipal wastewater services to the components of the water sector. Furthermore, central government started to be more deeply involved in the provision of municipal water and wastewater services.

The capital city, Amman, was facing challenges in water supply and wastewater disposal, services that had been rendered by the municipality. The water distribution service in the country, handled by the respective municipal councils, was not much better. Law number 1 for the year 1965 created a department in Amman Municipality to build and operate wastewater collection and treatment services. Another department of the municipality was empowered to care for its water projects and their operation and maintenance services by law number 14 for the year 1965 (Amman Water Law).

More autonomy was needed to care for these services, and the government responded by embarking on a more liberal water management policy. It established the Amman Water and Sewerage Authority (AWSA) by law number 19 for the year 1973 to care for the city's combined water and wastewater services. The law entrusted the new authority, AWSA, to seek and bring to Amman water from remote sources in cooperation with the NRA. For the rest of the country, the municipal water projects, heretofore assigned to the NRA, were shifted to a new agency, the Domestic Water Supply Corporation (DWSC), created by law number 56 for the year 1973. That law assigned to the new agency the responsibility of the distribution networks for the towns and villages throughout the kingdom, except Amman. For wastewater services elsewhere outside Amman, the Ministry of Municipal and Rural Affairs was in charge by virtue of another law, law number 12 for the year 1977.

This measure did not prove to be successful. It was more of a fragmentation of the administration of the water sector after its amalgamation in one central authority, CWA, and the successor, NRA. Reasons for the change were the frequent shortages of water supply to urban and rural areas and the inability of the NRA to cope with its wide-ranging responsibilities. Lack of finances to conduct feasibility studies and prepare projects for implementation was partly behind the delaying factors.

Policy toward the Jordan Valley

In the Jordan Valley, government policy shifted in regard to the philosophy of development that touched on water resources policy as well. The valley was a confrontational strip bordering Israel and the occupied West Bank. It suffered from paramilitary and military clashes with Israel after the June War. More than half of the houses in its villages were destroyed, and its population had dropped from 63,000 to about 5,000 people. When law and order were reestablished in 1971, the rehabilitation and development of the Jordan Valley became a high priority. The philosophy of development was transformed, and the government adopted an integrated social and economic development approach with irrigated agriculture as its backbone.

To that end, the government decided to assign the task to a new agency with a clear mandate and authority. The Jordan Valley Commission (JVC) was established under law number 2 for the year 1973. The JVC was to plan and implement the projects for the rehabilitation and development of the valley and hand them over, on completion, to the concerned government agencies for their operation and maintenance. The policy was amended in 1977, and the Jordan Valley Authority (JVA) was established by law number 18 for the year 1977 to succeed the Jordan Valley Commission, the Jordan River Tributaries Regional Corporation, and the parts of the Natural Resources Authority and the Domestic Water Supply Corporation operating in the Jordan Valley. Almost simultaneously, the valley region was expanded south to the Gulf of Aqaba on the Red Sea.

The law gave the JVA wide-ranging powers to develop and implement integrated social and economic plans for the Jordan Valley. The authority was concerned with not only water resources development and management, but also social development infrastructure, public utilities, transportation, housing, organization of farmers, municipal development responsibilities, and support for agricultural activities. It was empowered to seek finance for its projects from friendly countries and contract loans with the approval of the Council of Ministers. This drastic shift in policy proved to be successful and paid dividends in moving the Jordan Valley back onto the development trail and dramatically improving the standard of living of people there. (This is discussed further in Chapter 9.)

Transfer of Water to Amman

About the same time, Amman needed more water resources, and the first sizable transfer to the city from remote areas was started. At a high-level meeting chaired by the king, decisions were made to transfer water from Azraq,[7] some 72 km to the northeast of Amman, as well as the East Ghor Canal in the Jordan Valley. In absence of a dam on the Yarmouk, however, there still would not be enough water in the East Ghor Canal to meet agricultural demands in the valley and municipal demands in Amman, especially in the dry months. Irrigation water was rationed, and a formal decision was made to reuse treated wastewater for agricultural purposes in the Jordan Valley to partly compensate it for the water that would be pumped from the East Ghor Canal to Amman.

Shifts in Irrigation Technology

The JVC in 1973 reached a turning point in the policy of water resources development. It shifted the design of the conveyance systems of the East Ghor Canal from concrete-lined canals to pressure pipe networks. This change is credited to the persuasive abilities of World Bank professionals when they came to appraise one of four irrigation expansion projects in 1973.[8] The policy of the JVC was to make sprinkler lines available to farmers on a credit basis, and it had purchased enough lines to cover 9,300 ha. But this policy was rendered a failure when, on completion of the projects in 1979, farmers adopted more advanced irrigation methods, primarily drip systems. The shifts to advanced distribution networks and on-farm irrigation systems were prompted by the need to improve irrigation efficiency in light of the declining share of the Yarmouk waters caused by expanded Syrian uses upstream. Additional benefits of these advanced systems were increases in crop yields per unit water flow and unit land area.

Management of Municipal Water under Stress

The overall picture of municipal water supply and demand, and its forecast, had never been painted for the kingdom until 1978, when the National Planning Council, for lack of one water agency managing the municipal supplies for the whole country, commissioned a British firm, Howard Humphrey and Partners, to survey and forecast the municipal water demand for the northwest quadrant of Jordan, where more than 90% of the population lived.[9]

The results of the study shocked the officials of all water agencies. The water resources would not be sufficient to cope with the municipal and industrial demand of the year 2000 without infringing on the supplies meant to expand irrigated agriculture in the Jordan Valley. The shortfall was estimated at 160 mcm by the turn of the century, unless the kingdom took action immediately to make more supplies available.

These findings prompted the government to pressure Syria to approve the construction of the Maqarin Dam on the Yarmouk,[10] an international watercourse, and legislation was enacted in Jordan for its construction. The United States also enacted legislation to appropriate $150 million to finance the dam construction.

By 1980, however, and with Syria dragging its feet, relations between the two countries were strained over the Iraq–Iran War. Jordan sided with Iraq, and Syria sided with Iran. Jordan approached Iraq for the supply of municipal water, and Iraq agreed to have the supply gap in the year 2000 closed by transporting water from the Euphrates River to Amman. Howard Humphrey and Partners studied this project and found it to cost more per cubic meter of supply than the Jordanians could afford, so the project was shelved in 1984. Fortunately, groundwater was found in the Yarmouk Gorge and Wadi Arab that enabled the supply of municipal water to Irbid in the north and its environs.

The cities and villages in the country, including Amman, started to feel the pinch of municipal water shortages. The government and citizens were not pleased with the performance of the Domestic Water Supply Corporation and AWSA. Municipal supplies to Amman and later Irbid from the Jordan Valley water sources were entrusted to the Jordan Valley Authority in 1982 because of the urgency and based on the JVA's successful track record of project implementation.

Law number 34 for the year 1983 created a new, all-encompassing water authority known as Water Authority of Jordan (WAJ), which took over the distribution networks from the local councils all over the kingdom. WAJ was the legal successor of AWSA, the DWSC, and the parts of the NRA that worked on water resources. Differences over jurisdiction arose between the new authority and the JVA, and after a period of confusion that lasted for seven months, they were finally resolved. The two authorities henceforth remained in charge of the water sector, reminiscent of the administrative arrangement of the EGCA and CWA between 1959 and 1966. This dual arrangement was brought closer together in 1988 with the creation of the Ministry of Water and Irrigation. The two authorities remained autonomous, however, each supervised by a board of directors chaired by the minister of water and irrigation. Their respective laws were slightly amended to account for the new arrangements.

It had taken Jordan 17 years to realize that its arrangement for water administration initiated in the late 1950s had been more workable than any of the succeeding arrangements. However, the responsibility given to WAJ in 1983 to manage the distribution networks inside all the population centers in Jordan proved to be a burden. Rather, municipal water is better managed on the provincial or the city level, especially since the municipal councils have been reduced in number by merging several city councils neighboring each other into one. In Amman, the largest city, management of the operation and maintenance of water and wastewater systems was awarded in 2000 to a private sector joint venture of foreign and local companies, which has been running these services ever since.

This arrangement of management contracts involved suggestions from and conditions imposed by friendly donors regarding the water sector in Jordan. The World Bank was instrumental in promoting the management contract method as a step toward more advanced involvement of the private sector in water resources development and administration. But privatization of water resources, prematurely advocated by some donors, is a risky undertaking in Jordan under the prevailing social, economic, and political conditions.

Water Policy in the Peace Process

When the Middle East peace process started in October 1991, the minister of water and irrigation and his advisors foresaw the need for a crew of qualified professionals to start studying future alternatives for water supply in an era of peace. Bylaw number 54 for the year 1992 created a third branch in the ministry, operating under a secretary general.

The Water Agreement

In the peace process, an important policy was adopted by the Jordan government, formalized in 1993, and related to the negotiators. The water-sharing formula proposed by Eric Johnston, which had been accepted by the Arab League Technical Committee in September 1955 and distributed to the riparian parties in January 1956, was adopted. Jordan had to care for the East Bank share only; the negotiators were instructed to leave the West Bank share to the Palestine Liberation Organization (PLO) to negotiate over with Israel. This position was in line with the disengagement decision with the West Bank that Jordan's king had decreed in 1988.

Jordan successfully negotiated the water conflict with Israel on the basis of this policy and recovered whatever Jordanian share Israel was using (Chapter 10). The treaty of October 26, 1994, was made into law, which became effective on November 11. The treaty established the Joint Water Committee to handle all bilateral matters related to water and the implementation of the provisions of the Water Annex. The Jordanian side of the Joint Water Committee is selected from the staff of the Ministry of Water and Irrigation and is chaired by the secretary general of the Jordan Valley Authority. Coordination and cooperation between the two sides have been satisfactory to both of them. Improvements in joint undertakings in the water sector are expected after comprehensive peace is reached, in which case more riparian parties will join the cooperation process.

The Jordan Rift Valley and the Red Sea–Dead Sea Conduit

Another prominent feature of the peace treaty that went into law in both countries is the Jordan Rift Valley Project, modeled after the Jordanian experience in integrated development of the Jordan Valley. Like the water issue, the Jordan Rift Valley was addressed in a separate article in the treaty. Upon further talks under the auspices of the United States in the Trilateral Economic Committee (United States, Jordan, and Israel), the two parties came to an agreement to adopt the Jordanian Red Sea–Dead Sea linkage to mitigate the environmental impacts of the lowering of the Dead Sea level and to generate fresh water from the salty Red Sea water. The Jordan Rift Valley Project includes programs for integrated development in agriculture, transportation, energy, communication, electrification, public utilities, villages and urban areas, industries, and more.

A transfer of about 1,800 mcm of salt water is envisaged to raise the level of the Dead Sea to –396 meters below sea level, as compared with its historic level of –392 meters and its current level of –418 meters below sea level. The project was stalled for a time because of political reasons associated with the inclusion of Palestinians, but these difficulties were overcome in 2005 and efforts have resumed. A detailed feasibility study is expected to be awarded to qualified consultants, and fund-raising should begin shortly after the study is conducted.

Desalinated water is a direct outcome of the project. Red Sea water will be lifted about 220 m above sea level before it is allowed to flow by gravity to the Dead Sea. Thus the difference in elevation, or the total potential head, would be about 640 m, of which 550 m are needed to operate reverse osmosis systems of desalination; the rest of the total potential head will be used to generate electricity to help pump the desalinated water to consumption centers in Israel,

Palestine, and Jordan. The contribution of this project to Jordan water resources will depend on the rate of flow and the agreement on sharing the water, which presumably would be in proportion to the evaporation of the Dead Sea emanating from the territorial jurisdiction of each party.

Formulation of Water Strategy and Policies

A formal water strategy for the country was developed by the minister of water and irrigation in April 1997 and endorsed by the Council of Ministers on May 3 (MWI 1997) as the official strategy of the government. Under it, four policies were formulated and the documents endorsed by the council.[11] The water strategy set goals for the long term, and the policies were created to achieve the goals over time. While the strategy is more fixed, the policies respond to changing needs and are amended accordingly.

Perhaps the most important factor that dictated the water strategy objectives and policies was the gross imbalance in the population–water resources equation, which has shown huge deficits. Goals were set to maximize efficiency of water conveyance, distribution, and use and to augment available water resources through brackish water desalination and treated wastewater reuse. Both objectives were set through decisions of the Council of Ministers in accordance with the laws. An increase in efficiency has been promoted by measures of supply and of demand management, including adoption of advanced systems in irrigation and the replacement of old networks. Agricultural credit, run by a government institution, was extended to promote advanced irrigation systems. Demand management measures were taken through upward adjustment of water charges.

As things stand today, one institution in the country, the Ministry of Water and Irrigation, is in charge of water resources development, distribution, and management, with the exception of green water.

The role that friendly donors have been playing in the water sector of Jordan cannot be ignored. The World Bank was the first to draw attention to the need of a policy for cost recovery as far back as the early 1960s, when it helped finance, with International Development Association (IDA) funds, water projects in the kingdom, such as the Beit Sahur–Jerusalem water project. The bank made recovery of operation and maintenance costs and a portion of the capital cost a condition for effectiveness of an IDA loan to Jordan to finance the Northeast Ghor Irrigation Project in 1974. The government responded by stating a policy then aiming at recovering the full operation and maintenance costs of irrigation and a portion of the capital cost. However, irrigation cost recovery has always been a problem. (Attempts to reach that goal are described in Chapter 6.)

All other donors followed suit, among them the Kreditanstalt für Wiederaufbau (KFW) of Germany, which refused to extend further loans to the municipal water projects in 1997. The government adjusted the water tariff to recover the full operation and maintenance costs, and KFW then extended two loans: one to rehabilitate a portion of the Amman municipal water network, and another to double the capacity of water transmission from the Zai plant to the Daboug Reservoir in Amman.

Possible Future Legislation and Administration

Looking ahead, Jordan needs to legislate for the management of its green-water resources; its fossil water, which is not renewable; its gray-water resources, which will play a more important role; and the management of its international and transboundary waters.

Green-Water Management

Water resources in the country are public property by legislation. Green water forms a substantial percentage of the country's water resources, and it is almost equal in potential to the indigenous blue-water resources (Chapter 7). Because water resources are public property, the important component of green water should receive the management care that other components receive.

The challenge in management of green water is the location of its availability. It is stored in the soil of agricultural lands, which are mainly privately managed and owned. Actually, landownership in Jordan is of two classes: one pertaining to the land parcels inside the municipal boundaries of cities, towns, and villages, and the other the land outside those boundaries. Ownership of the first is absolute and is called *mulk;* in the second case, it is less than absolute, classified as *miri* (belonging to the Emir on behalf of the public), but the "owner" of such property, called a *mutasarrif,* is free to use it in accordance with the applicable laws. Until recently, the inheritance law regarding the first type gave a male heir twice the share of a female, but regarding the second type, the same law gave equal shares to male and female heirs.

The ownership classification of agricultural land as *miri* includes a window for a public–private partnership in its utilization. The *mutasarrif* is interested in cultivating the land, and the public, represented by the state, is interested in preserving the water resources that the land stores (green water) and maximizing the use of that water. Any green water that is not used will be lost to evaporation during the hot, dry months of the year, and this resource is especially needed in a water-strained country like Jordan; hence, this is a justification for public involvement.

Agricultural land farming is not always secure because of the interannual variability of rainfall. A farmer would have to take a risk when sowing the seeds for winter crops in November if good showers have not occurred by then. In this regard, government can make an educated forecast and evaluate the risk. It can go into partnership with the farmer to contribute to the cost of production and carry part of the risk. If the farmer recovers the cost of production, so would the government, and if the farmer made a profit, the government would take a token share of it. In fact, government has become involved in similar arrangements in irrigated agriculture in the Jordan Valley. The JVA, a government organization, is responsible for the supply of water shares to the farmers. When the irrigation water supply is not quite secure, the government pays farmers to leave their farms in the project area fallow to avoid providing them with water that is needed for other purposes.

Urbanization has infringed upon agricultural land at rates that should not be acceptable in some areas, such as Irbid and environs in the north; Sweileh, Baqa'a, Salt, Amman and environs, and Madaba in the central region; and Qasr, Rabbah, and Karak in the south. The soil has been replaced with concrete and asphalt, the recharge areas of aquifers have shrunk, and the green water is lost except for whatever gardening is practiced inside these conurbations.

In the case of green-water preservation and use, the involvement of government in its management should be based on legislation that regulates the relationship between government and the landowners and croppers and defines responsibilities.

Fossil-Water Management

Fossil water underlies the entire country, with the exception of the igneous rock stretch north of Aqaba. This water is historic and not renewable. It is regarded as a strategic reserve that should be used sparingly and for crucial purposes only. Dependence on it for long periods of time carries a high risk of depletion, as well as serious environmental, social, and security hazards.

Legislation is needed to carefully manage this resource and allocate it to purposes that are clearly in the public good. In 1983, the government was encouraged by the results of exploration for groundwater in the south, at Disi-Mudawwara and Sahl As Suwwan, and pioneered the use of the fossil water in that area for agricultural purposes. It had several wells drilled and started out the farming activity in 1984 with winter wheat.[12] The idea originated during the frequent trips that the king, accompanied by the prime minister and other officials, was making to Saudi Arabia. They flew over the irrigated wheat fields there and wondered why such an undertaking could not be emulated in Jordan. Another motivation was the preservation of Jordan's rights to the fossil aquifer that also extends under Saudi Arabia.

Soon after the pioneering effort was initiated, it became clear that the government could not continue in the business of agricultural production. The investment in the well fields, electrical transmission to them from the national grid, and some machinery had already been made, and there was no conceivable way that such investments could be recovered through government farming, nor would the government be able to sustain the wheat production activities. Actually, the JVA had already gone through a sad experience trying to pioneer agricultural production with advanced techniques in Wadi Araba, a region very similar in climate and population to Disi. It became clear in 1981 that the JVA, although an autonomous agency that was remote from government routine, could not sustain agricultural production beyond 1983. Governments simply are not farmers.

To recover the investment, or part thereof, and keep the momentum going, the government announced its intention to permit private companies to exploit the aquifer for wheat plantation. (Wheat is heavily imported, with domestic production accounting for only about 15% of the country's bread needs.) Companies were hurriedly formed, and they rented the state land for nominal charges, paid for the assets, and replaced the government in caring for agricultural production. A change was made to the cropping pattern, and soon those companies started to produce potatoes and watermelons that competed with the Jordan Valley crops. The amount of water abstracted annually reached about 85 mcm, but the government, motivated by the World Bank through the Agricultural Sector Adjustment Loan, demanded a phased reduction. Little was achieved in reducing the abstracted amounts, however. More recently, the Ministry of Water and Irrigation imposed charges on water abstracted in excess of a permitted amount for each well, and that measure upset the owners of the companies. They went to court, and the verdict was in favor of the ministry, as the decision was taken in accordance with new legislation.

In 2005, a new government was formed, and it reversed the policy of charging for the fossil water on the grounds that only the contracts signed with the companies regulate their relationship, and those contracts have no mention of water charges, only land rentals. Legally, the new government and the companies are right. The Supreme Court mandate is to rule over management decisions and not on issues of contract disputes. It rightfully ruled that the decision of the ministry to charge for water is legal and is in line with the regulation that was issued to control groundwater resources. The deficiency was in the regulation that ignored acquired rights, a matter to be ruled by a different court, the Court of First Instance.

Fossil water for agricultural development should be curtailed to a minimum, and the economic and social benefits of its use weighed against the environmental, social, and political risks. In short, legislation is needed to regulate the management of such a strategic reserve.

Gray-Water Management

Treated wastewater has become an important water resource, used primarily in agriculture. Some of it, treated locally in cesspools, is recharging groundwater aquifers used for both agriculture and municipal purposes. The treated waste-water flow is on the rise, and its use has become a familiar option.

The wastewater management policy as approved by the government speci-fies measures that are to be taken to make safe the use of treated wastewater. However, legislation clearly aiming at regulating the quality of effluents is urgently needed. Moreover, the drive to expand wastewater collection and treat-ment networks should increase its momentum, especially in towns and cities whose cesspools are known to contribute to groundwater recharge. The Amman–Zarqa conurbation is a prime candidate to become fully served and connected to wastewater collection networks and treatment facilities.

International and Transboundary Water Management

International water refers to shared surface-water flows, whereas transbound-ary water refers to groundwater that traverses the boundaries between any two countries. Another crucial need for legislation is the management of these shared waters. This is normally done through treaties that regulate the rights and obligations of the contracting parties. Two treaties have been enacted into law: one with a detailed Water Annex between Jordan and Israel, which has been fairly managed so far, and another with Syria for the utilization of the Yarmouk River. Legislation is still needed to follow up on the implementation of both existing treaties, to elaborate the tasks of the Jordanian side in each, and attend to the details of implementation.

Additionally, sharing the groundwater underlying Syria and Jordan, and especially that feeding the Yarmouk base flow, is not regulated. This fact adversely affects Jordan in the Yarmouk Basin, where the kingdom's share of the spring flow has been diminishing as a result of Syrian exploitation of the aquifer feeding them. The Yarmouk flow in August 1963 at Adassiyya in Jordan, for example, was 6.3 m^3/sec. In August 2004, the flow did not amount to more than 1.8 m^3/sec. Other adverse effects on Jordan are manifested in the Yarmouk groundwater basin and the rate of inflow of transboundary water from Syria. An agreement with Syria on the regulation of groundwater uses is clearly needed.

Water concerns exist with other neighbors as well, primarily Saudi Arabia to the south and east and Iraq to the east. Groundwater underlies the territories of Jordan and these countries, and the connected aquifers contain primarily fossil water. Treaties with these neighbors are needed to regulate the exploita-tion of these aquifers.

In May 1997, the UN General Assembly adopted a convention on the Non-navigational Uses of International Watercourses. Jordan was among the first states in the world that ratified the convention, but member states have been slow in ratifying it. By 2004, only 17 countries have done so. The provisions of this convention would regulate the relationships between riparian countries in managing the surface water and groundwater of a shared basin. However, Jordan's existing agreements over water resources with Israel and Syria were concluded before the UN adopted the convention, so it had no bearing on the outcome of negotiations between Jordan and each of its coriparian parties.

Summary

Jordan went through several stages in the formulation of its water policy. At the outset, it was heavily influenced by exogenous factors related to the Jordan River Basin. The ideas of foreign advisors and consultants who served the kingdom in 1939, 1949–1951, and 1953–1955 were instrumental in formulating policy. A rather stable formula for the sharing of Jordan Basin waters was worked out by a U.S. presidential envoy, Eric Johnston, who negotiated the formula with Arab and Israeli parties between 1953 and 1955. The importance Jordan placed on its share of international waters prompted it to conclude agreements with Syria for the joint utilization of the shared river, the Yarmouk.

Jordan adopted a water policy to serve economic and social development and enacted several pieces of legislation to that end. Most notable was the policy adopted for the development of the Jordan Valley and the central management of water resources outside the valley. Still, regional developments affected Jordan's water policy, particularly the implementation of Israel's plan to unilaterally divert Jordan River waters to the Negev.

The policy oscillated between central management and pluralism of water administration, finally settling on a central administration. The competition with Israel over water resources was finally resolved through bilateral negotiations conducted under the Middle East peace process. Subsequent to the settlement of the water dispute, Jordan embarked on the official formulation of a water strategy and a set of policies pertaining to irrigation, municipal and industrial water, groundwater, and wastewater management.

It is important for Jordan to consider the future use of its water resources and provide appropriate legislation for the management of green water, fossil water, and gray water, as well as international and transboundary waters.

References

Baker, M., Jr., Inc., and Harza Engineering. 1955. *Yarmouk–Jordan Valley Project: Master Plan Report*. Amman, Jordan: Ministry of Finance, submitted to Jordan–United States Technical Services.

Dokhgan, O. 2004. Personal communication between Omar Abdullah Dokhgan and the author, February 20.

Haddadin, M. 2001. *Diplomacy on the Jordan: International Conflict and Negotiated Resolution*. Norwell, MA: Kluwer Academic Publishers.

Ionides, M. 1946. Jordan Valley Irrigation in Transjordan. *Engineering.* September 13, 1946, reporting on a study prepared in 1939.

MacDonald, Sir Murdoch, and Partners. 1951. *Report on the Proposed Extension of Irrigation in the Jordan Valley.* London: Sir Murdoch MacDonald and Partners.

MOFA (Minister of Foreign Affairs). 1958. Memorandum no. 58/14/6719 from the foreign minister of Jordan to the U.S. chargé d'affaires in Amman. February 25, 1958.

MWI (Ministry of Water and Irrigation). 1997. *Jordan's Water Strategy.* Article no. 10. Amman, Jordan: MWI.

Notes

1. For details see Haddadin 2001.

2. In April 1957, the king of Jordan aborted an attempted military coup; in February 1958, Egypt united with Syria, and the new state, the United Arab Republic, targeted Jordan for political upheavals; in July of the same year, a military coup abolished the Hashemite monarch in Iraq and massacred the royal family.

3. Nicholi Simansky was first recruited from the Sudan by G.F. Welpole, the British director of lands and surveys in Jordan.

4. The law submitted to Parliament by the Council of Ministers stipulated that the director of the Central Water Authority had to be a foreign national, indicating the influence of USAID on water policy details. In spite of the deletion of that provision from the law, an American engineer was the first director in 1959.

5. Attempts were made by the United States in the years following Johnston's shuttle diplomacy to have all the riparian parties endorse the Unified Jordan Valley Plan. No such consensus was reached, and Israel proceeded, with the United States' prior knowledge, tacit consent, and indirect finance, to build its National Water Carrier from northwest Lake Tiberias to the Negev. See Haddadin 2001 for details.

6. A major component of the project was the Mukheiba Dam on the Yarmouk, straddling the Jordanian and Syrian banks. Other components included conveyor canals meant to replace water drawn from the Jordan River to irrigate Jordanian lands in its floodplain (about 6,800 ha) with irrigation water from the Mukheiba Dam.

7. The transfer was first decided to come from the King Talal Dam, some 40 km to the northwest of Amman. I was adamantly against that project because of a potential pollution problem and succeeded in stopping it when it was reviewed just before the construction contract was to be awarded in a meeting at the National Planning Council.

8. Team leader Guy Lemoigne spent hours elaborating the advantages of pressure pipe networks.

9. The region delineated by the National Planning Council lies between the Wadi Wala in the south and the Yarmouk River in the north, and from the railroad to the east and the Jordan Valley to the west.

10. The Maqarin Dam would straddle territories inside Jordan and Syria and inundate regions of both countries. An agreement was reached for cooperation between the two

countries for that objective in 1953. Syria's approval was key to start the construction of that dam.

11. Water Utility Policy, Irrigation Water Policy, Groundwater Management Policy, and Wastewater Management Policy were successively formulated and issued by the Council of Ministers in 1997–1998. They are available at the Ministry of Water and Irrigation, Amman, Jordan.

12. When that decision was made in 1983, I was the only one among the 16 members of the Higher Agricultural Council, chaired by the prime minister, to express reservations about such an undertaking, on the grounds that fossil water is too valuable to exploit for agricultural production. I was excluded from any involvement in the project because of my objection.

3

Water Data Systems in Support of Water Management and Planning

Hala Zawati Katkhuda and

Munther J. Haddadin

Water data are as essential to project planning as water itself is essential to all forms of life. Projects and their management are the vehicle for implementation of a meaningful water policy. Water projects and management are well planned and executed only if availability of data is ascertained, and as data become readily available, the implementation of water policy has better chances of success. Additionally, data gathered in the implementation phase of water policy are employed as a feedback input into formulating and updating the water policy, or for injecting measures for a midcourse correction of certain aspects of the policy.

Essential catalysts to enhance the central role of data have emerged. Water data systems have become increasingly helpful in data collection, processing, analysis, and dissemination. They thus have assumed an increasingly important role in water resource assessment, development, management, and update. With the proliferation of digital computation, modeling, and techniques supplied by information technology, water data systems have enhanced the role that data have been playing in water resource development and management. The data models involve a series of steps in what can be termed as water data management (WDM), and in most cases these steps form an iterative process.

WDM consists of data monitoring, documenting (in the various stages), analysis, and assessment of the impact of the data on the envisioned projects and ultimately on the policy itself. In light of data processing and evaluation, amendments to the policy may be warranted, thereby the iterative nature of the interaction between data management and water policy.

Jordan, like several other countries in the region and developing countries at large, was not quick to benefit from modern technology, nor did it have the know-how to embark upon adequate data management as technological innovations became accessible. This was a handicap that necessitated the use of data synthesis techniques in absence of much-needed actual, properly monitored data. Stations for measurement of rainfall, wind speed, and ambient temperature, for example, were thought of as early as 1937 (MWI 2004). Data collected prior to the proclamation of the kingdom in 1946 were either little or nonexistent.[1]

In recent years, the situation changed, and Jordan has come to recognize the importance of data monitoring, compilation, and analysis in facilitating policy formulation, evaluation, and implementation. The provision of a comprehensive water data–monitoring system, fully integrated with an adequate data and information management system, has surfaced and gained priority in water resource management. Government policy took a welcome turn toward endorsing such an important priority and authorized the minister of planning, in cooperation with the minister of water and irrigation, to facilitate the realization of that priority. The two integrated water systems, monitoring and information, are needed to organize data collection, verification, management, dissemination, and analysis. The systems are essential for all water management operations and constitute the first step for water policy formulation and implementation.

This chapter describes the efforts made to implement government policy to provide water data systems in Jordan, as well as the impact the availability of these systems had on water policy formulation in the country.

Jordanian Water-Monitoring Networks

A well-conceived, comprehensive, and integrated water quantity–quality monitoring program is essential and critical to plan, formulate, and implement water policies; evaluate the anticipated benefits; and assess the effects of their implementation. All these steps are only as accurate and reliable as the data used to determine them. It is therefore clear that water policy planning, design, implementation, and evaluation should always be founded on accurate and reliable data and on appropriate statistical methods and analytical tools of forecast and prediction.

The Jordan government arm involved with the constitutional and legal responsibilities surrounding the administration of the water sector has been, since 1989, the Ministry of Water and Irrigation (MWI), the successor of several previous institutions (see Chapter 2). Each has given attention, mostly with the help of friendly donor agencies, to the introduction, development, and upgrading of the water quality–quantity monitoring networks. The first monitoring stations were rainfall stations and were installed at the time of the British Mandate in 1937 in Ajloun and Mafraq governorates. Stream flow–gauging stations were first installed on the Yarmouk and its tributaries in 1952 as part of the study of the river sponsored by the United States and the United Nations Relief and Works Agency for Palestine Refugees (UNRWA) to build the Bunger Dam (Chapter 10). Flow measurement of the Yarmouk was basic to the planning for the use of that river in the development of the Jordan Valley, to which the government assigned a high priority. Other stream flow–gauging stations replaced the older ones on the Yarmouk at Maqarin and Adassiyya in 1963, in the course of studying the feasibility of a dam at Mukheiba (MWI 2004).

Since then, the predecessors of the MWI, especially the Central Water Authority (1959–1965) and the Natural Resources Authority (1965–1984), and then the ministry itself have continuously provided, managed, and upgraded the water-gauging network throughout Jordan.

In its constant quest for improvement of its monitoring and data management and data systems, MWI laid down, for the first time, an official water sector strategy for the country in 1997. The water strategy was discussed by the boards of the Water Authority of Jordan and the Jordan Valley Authority, the two wings of MWI. The draft water strategy, after it was debated by the boards, was discussed with experts working for the World Bank in the Middle East and North Africa (MENA) Department.[2] With the draft thus finalized, the minister of water and irrigation forwarded it to the Council of Ministers for adoption as a formal government strategy to be observed by all concerned.[3] The strategy was passed in 1997 by the Council of Ministers as transmitted to the minister of water via letter number 57-11-1-3580, dated May 3, 1997. The water strategy defined the long-term goals that the government of Jordan sought to achieve in the water sector, among which was the support of a comprehensive national water databank that would be well supported by a program and system for monitoring. In this regard, the strategy stipulates:

> Water is a national resource and shall be valued as such at all times. A comprehensive national water databank will be established and kept at the Ministry of Water and Irrigation, and shall be supported by a decision support system. It will be supported by a program of monitoring and a system of data collection, entry, updating, processing, and dissemination of information, and will be designed to become a terminal in a regional data bank setup. (MWI 1997)

Five years later, in 2002, the water data situation was very promising. Jordan, represented by the Ministry of Water and Irrigation, had covered most important water-monitoring areas. The number of monitoring stations had increased to 220 rainfall stations; 44 gauging stations, along with discharge measurements of around 500 springs; 34 evaporation stations; and 117 water level–monitoring wells (MWI 2002). In parallel with these efforts, a comprehensive and integrated program to monitor water quality in all basins, from all sources, and for all uses was implemented. Some stations were old and others were new, employing modern technologies. As of 2006, the upgrading process of these stations has slowed down because of financial or technical constraints. The latest technologies are usually more sophisticated, requiring upgraded technical skills for maintenance that developing countries need to invest in, and the costs of operation and maintenance are beyond what the tight government budgets allow.

In 2003, Jordan adopted telemetry technology, installing telemetry online monitoring stations in two pilot projects in cooperation with friendly donor agencies.[4] These stations use the latest technologies in monitoring water quality, quantity, and meteorological parameters, supplying online data to decisionmakers, a drastic departure from conventional water management techniques that had been practiced earlier. If these two pilot projects prove to be successful through competence in indigenous management, use, and proper maintenance, further acquisition of similar stations will most likely receive donors' support. Replacement of older stations employing outdated technologies becomes attractive. These telemetry stations automatically take measurements of parameters and transmit data immediately to a data center, bypassing the traditional steps for data collection and transfer, including manual data entry, which consume time and effort and increase the likelihood of errors. In addition, these stations make the data available to decisionmakers online.

Data Collection, Verification, and Availability

Collected water data can be divided into two categories: water resource data and water use data.

Water Resource Data

Water resource data are collected on a regular basis from monitoring stations and sites; this category includes data on water quality and quantity, as well as meteorological data.

The main official government agency responsible for water quality in Jordan is the Ministry of Water and Irrigation; its job here is to ensure that the quality of water provided meets the standards for the intended use. National standards and specifications have been set for water used for different purposes in the country.[5] Water quality data are available in the Ministry of Water and Irrigation laboratories.[6] Law number 54 for the year 2002, as did its predecessor, made the Ministry of Health the ultimate authority on the fitness of water to be pumped into the domestic water networks. The Ministry of Health has its own laboratories for monitoring domestic water quality. Other organizations, such as in some municipalities in Jordan, are also responsible for testing water samples from specific sites and therefore have their own laboratories and sets of data. Some nongovernmental organizations (NGOs), such as the Royal Scientific Society, have their own labs to monitor quality; others, such as the Royal Society for the Conservation of Nature, outsource the testing tasks. Most universities with water or environmental programs maintain laboratories covering some or most parameters investigated in water testing.

Data on water resources, including groundwater and surface water, in addition to meteorological data, are collected on daily, monthly, quarterly, seasonal, or yearly bases, depending on the type of station installed and resource measured. Currently the data, with the exception of the pilot projects above, are collected and transferred manually from stations distributed all over the kingdom; some data are recorded automatically, and some manually in the field. Recorded charts are then analyzed, organized in tables, and sent to the data section of the Ministry of Water and Irrigation, called the Water Information System (WIS).

Verification procedures of collected data or data entered into the system are still limited. This raises many issues regarding the quality of data entered into the system. In addition, historical data in the system need verification to correct inaccuracies therein and minimize the uncertainty associated with them. However, according to former water officials in charge of data collection and management, focus on the needs of the data collection divisions started to decline particularly since the June war of 1967. Accessibility to water resources, usually in areas remote from the capital, required transportation, logistics, support, and protection for security, especially during the four years of turbulence that followed the war. Such focus gradually eroded because the government's attention was focused on security, stability, and law enforcement. It was not uncommon for armed militias to stop water officials on the road and make their mission difficult. Moreover, the successive changes in the institutional structure that dealt with the water sector (see Chapter 2) made more urgent priorities emerge. In 1997, for example, the logistics accorded to the data section in the MWI were so meager and the vehicles were so run-down that field operations in data collection and monitoring became a difficult if not impossible task.

Although the WIS has a wealth of data, some are not very reliable and need verification before use. The managers of the system are constantly laboring to check, validate, compare, and improve the quality of data in the system.

Still, there exist some handicaps in the data management process today. Standard operating procedures for data measurements and management have not been adopted. No written procedures are available for data collection, verification, and documentation, nor are there any quality assurance measures of data entry. Adopting and enforcing accepted procedures and employing measures for data quality assurance will increase the reliability of data. The adequacy of these procedures, or lack thereof, will directly influence the validity, appropriateness, and adequacy of the projects planned in their accordance and thus will play a role in the success or failure of policies formulated on the basis of such data. The good news, however, is that water data management personnel are aware of these handicaps and are making serious efforts to overcome them. Such a state of affairs is not the result of a lack of policy, but can be blamed on the financial appropriations having been cut back by virtue of the need to minimize government budget deficits.

Water Use Data

Data collected on water uses in Jordan have been synthesized in many cases. Some computations of consumption depend on actual measurements (part of municipal and industrial uses), and others are pure estimates; ready examples are current and old data on pumping for irrigation from private wells. These estimates entail a level of inaccuracy in the data and can have wide margins of error.

The new bylaw number 85 adopted in Jordan in 2002 will help in improving the quality of these sets of data.[7] It aims to control the amounts of water extracted from private wells. Article 29A of this bylaw stipulates:

> Every owner of a well drilled and tested in accordance with the provisions of this By-Law should obtain before commencement of utilization thereof a license for water extraction issued by the Secretary General or his designee specifying the conditions that the licensee should comply with, including the following:
> 1. The maximum amount of water that may be extracted from the well within a fixed period of time.
> 2. The purpose of water use.
> 3. The maximum area that may be irrigated from the water of the well licensed for agricultural purposes.
> 4. The installation, at the expense of the owner of the well, of a water meter after it has been approved and sealed by the Authority. This condition should be complied with prior to the issuance of water extraction license.

The article goes on to enumerate other obligations. Enforcing the installation of water meters on private wells, as in item number 4, makes data collected on amounts of water extracted from private wells measurable, therefore reducing estimated values and increasing accuracy. The bylaw was an outcome of a change in policy, and enforcement of the new policy has affected the quality of data already. The improvement of data quality will reflect on the statistics and reports produced from the WIS for policy reevaluation, amendment, and implementation. This serves as a good example of the interactive nature of the iterative process between data management and policy.

On the other hand, some systems for data collection of water uses, distribution, and management in Jordan are highly sophisticated and accurate. The system used by the Jordan Valley Authority for the distribution of water in the valley is a case in point. It includes a System Control and Data Acquisition (SCADA) System (Grawitz and Hassan n.d.) and was installed with the help of yet another donor agency in 1995.[8] The data are captured into a management information system created specifically for this purpose. The system covers all tasks related to irrigation water management in the valley, from the measurement of water at the source before it enters the Jordan Valley irrigation infrastructure to water distribution to farmers, including billing and accounting. It helps set and define the water management policy for irrigation in the valley in any given year; this serves as an example of the dependency of the water policy on the output of the SCADA system.

Thus data collection and verification mechanisms for water uses in Jordan include accurate automated measurements, manual measurements, and in quite a number of cases, pure estimates.

Development and Availability of Data Systems and Tools

Policy analysis and implementation are as adequate as the data on which they are based. Data represents facts, which are, by definition, accurate and reliable. Easy access to them is much facilitated through a databanking system. Furthermore, data analysis should be automated through the employment of user-friendly analytical, predictive, and forecasting tools fully integrated with the data systems.

In its search for developing and upgrading systems and tools, Jordan has gone through many phases. Efforts began in the early 1980s, when the accumulation of piles of water data led to the procurement of hardware and the development of a software application system to store, organize, and archive the data. The system was primitive, but it served the main purpose: gathering of data into one system where it can be saved, preserved, and archived. That marked the event when water data management began knocking on the door of modernity and denoted a major shift in the policy of water data management.

By the early 1990s, the system proved not to be adequate. Decisionmakers, policymakers, and planners, having tasted the fruits of automation, started asking for more. An interactive system became essential, and analysis and reporting tools were important to incorporate. A new system was developed; it was interactive and user-friendly, and it included data for all water resources and part of the water uses in Jordan. Static as well as time-series data were incorporated.

The system represented a major step forward. Automated queries by users were accommodated, an innovation of the time; reporting facilities were introduced; and annual and monthly publications were automated and easily produced. All this represented a major improvement in the presentation of facts to policymakers, which in turn improved and accelerated policy analysis and evaluation. The system was called the Water Data Bank of Jordan and was implemented in close cooperation between the Jordanian government and the United Nations Development Programme (UNDP).[9] The system represented a step forward, and although it was viewed as a giant leap by users and junior workers, many more steps were still to follow.

In 1995, Jordan improved the water databank system and included new data types and reporting facilities with the help of another friendly donor.[10] Fine-tuning, upgrading, and updating of this system have been ongoing. In the process, the responsibility for water data management was shifted away from the Water Authority and onto the ministry, taking advantage of the attention accorded to the ministry branch that was initiated in 1993. The resulting updated system is the Water Information System (WIS), managed by the Ministry of Water and Irrigation.

After the creation of the WIS, which hosted most of the data important for planning and management, it was important to integrate this system with analytical and predictive tools that would facilitate proper planning and improve management. Some such tools were integrated with the WIS; others are stand-alone tools. The following tools were developed, deployed, and are in use:

- Geographical information system (GIS)–based tools with algorithms used for National Water Master Planning (NWMP).[11] Started in 1997, the project developed the database tools necessary to manage water data. It provides water specialists with data and information for water sector monitoring, planning, and management. The new Digital Water Master Plan is composed of digital planning software tools that are based on the WIS available at the Ministry of Water and Irrigation. These software tools are database applications with a GIS interface that are used to do the following:
  - assess the present availability, withdrawals, losses and uses of the water resources;
  - formulate alternative development scenarios for water resources and demand and use at various planning horizons;
  - perform the balance of resource availability versus actual or anticipated demands; and
  - identify technical and operational options in order to bridge the gap between resource availability and demand. (MWI 2003)
- GIS-based tools to assist in water resource management, land use and management, water distribution and supply, and simple maps with thematic

presentation. These systems are still in their initial phases; a lot can be done with GIS to facilitate and simplify water policy decisions and planning. Visual geographic presentations along with the analysis a GIS can provide are important benefits for water professionals.

- The Water Management Information System (WMIS) connected to a SCADA system. This is another tool used at the Jordan Valley Authority to help achieve optimal management of water resources and distribution (Grawitz and Hassan n.d.).
- A Laboratory Information Management System, customized to manage the water quality data. This system provides tools for lab management and data dissemination. It represents an excellent database for historic water quality information, essential for policy design and assessment of policy implementation.

Data Publication, Dissemination, and Accessibility

Water data collection and entry into data systems should always be followed by dissemination and publication of the data and information. Ease of accessing the data, whether in publications or online, is essential for water data seekers, such as researchers, planners, and students. Absence of proper publications and data dissemination procedures makes it harder for both data seekers and analysts to do their job. Formal government policies to this end have not been formulated, but they need to be.

In many countries of the semiarid region of the Middle East, in light of competition over shared water resources, water data availability has been viewed by water officials as sensitive information that should be classified. Official government policies in this regard were never stipulated or announced. In the case of Jordan, other factors came into play. One was a quick turnover of ministers with some lack of acquaintance with the water sector, thereby coming under the influence of their professional subordinates. Another was the general belief that data equate to power, and the officials in charge of data preferred to keep them close to their chests. A third factor was job security, which was enhanced if the people in charge of data were singular in performing the job, keeping even their colleagues in the dark. A fourth was the self-satisfaction that accrues to employees who know they are always wanted and almost indispensable.

This attitude has been gradually changing, especially after concluding peace with Israel and ending competition with it over water resources. The situation with Syria also is improving in light of the implementation of the Wehda Dam, and the water officials are less inhibited about releasing information.

Perhaps the only inhibition still affecting dissemination of specific data is the questionable reliability of data that are unrevealed or still raw, unchecked, or unanalyzed. Certain data are purely absent and nonexistent.

During the Middle East peace process, the chief water negotiator of Jordan advocated a policy of transparency with respect to data, and his position was endorsed by the High Committee on Negotiations. He contended that transparency regarding the water sector helps Jordan a lot, being a water-poor country and belonging to the lower-middle income category. He pointed out that Jordan's challenges could be expressed in a single word: *liquidity*. The country is poor in water, oil, and financial resources. Being transparent with data concerning them can help attract assistance to overcome these challenges. The water strategy paper (MWI 1997) adopted by the Council of Ministers points to the openness of Jordan in this regard and aims at making the water databank of Jordan part of a regional network in which parties can share the information.

Publishing data on a monthly or yearly basis should be an easy task. The process is automated when the data are available in a database system. Publishing tools should be integrated with the data management system to allow fast and accurate publication. Water publications are done by different departments, however, and there is no standard unified system for publications; some departments have tools integrated with their database systems, but others put together their publications manually and thus spend a lot of time and effort, in addition to increasing the possibility of errors.

The Ministry of Water and Irrigation, in its WIS, hosts most of the data concerning water monitoring, management, and planning. The WIS is the sole official authorized source for water data to external and internal users. It is a serious disadvantage that no publications are currently issued from this system, although tools are set and ready to publish various types of data, such as rainfall (averages, frequency, intensity), evaporation, stream flow, springs discharges, water quality, groundwater levels, well production, and the effluent and influent of wastewater treatment plants (MWI 2004). Available publications from the system date to 1994 and earlier.[12]

Accessibility to data continues to be hindered, for the same reasons as indicated above. There is no official policy to hide data from data seekers; to the contrary, the water strategy stipulates otherwise. Improvements in data accessibility and dissemination are hoped for as efforts intensify to upgrade the WIS and the capabilities thereof. But data dissemination and accessibility policies are not in the best of shape. They are not clear; no written procedures are available that define who can access what and how. In the absence of transparency, personal and biased judgments that tend to withhold information at times may prevail, and this definitely is not what ministry officials want.

Potentials for Improvement

During the past twenty years, Jordan gave priority to the task of developing a water data management system. The efforts were supported by many international donor agencies, including the United States Agency for International Development (USAID), UNDP, Gesellschaft für Technische Zusammenarbeit (GTZ), Japan International Cooperation Agency (JICA), the European Union (EU), and Kreditanstalt für Wiederaufbau (KFW). They shared with the Ministry of Water and Irrigation and its predecessors the belief in the need for a water data management system to guide management of the water sector in Jordan.

The system and tools that have been developed so far are good, but they still need enhancement through such things as tools that serve the decisionmakers, systems integration, and unification. The system must serve more users, including professionals, researchers, and students, who currently face barriers in their attempts to access it. It must also be easily accessed by decisionmakers, auditors, and even the press. There is nothing wrong with posting the processed information on a website.

The main aim of improvements is to come up with a unified national water data management system that incorporates all water data types and all tools needed to facilitate the management of the stressed water sector with relative ease. To achieve this goal, the following improvements are proposed:

- All water data types available under different departments and organizations must be gathered together in one unified data system that applies standard operating procedures for the monitoring of parameters, data collection, verification, entry, and dissemination. This can be a decentralized system in which every organization manages its own data, but the data can be accessed by anyone through this unified, shared system.
- Quality assurance procedures should be employed to verify both historical and newly entered data.
- A standard policy for data access and dissemination must be formulated, clearly delineating open versus classified files.
- Tools developed or adopted for data analysis, modeling, graphic, and geographical representation should always be integrated into the database system; stand-alone tools are far less efficient in providing updated analysis and information to decisionmakers and planners. They also require a lot of effort and time to run.
- Unification of coding systems used for identifying resources is mandatory. Much effort needs to be exerted to unify the different systems currently used in different organizations. There can never be a unified data management system if the coding system is not unified.
- Available reporting facilities must be used and enhanced. Monthly and yearly publications for many types of data gathered in the system are necessary. Tools available for publications must be used, and other needed tools developed. Substantial amounts of time and effort have been exerted to gather these data from different sources. Failing to publish them monthly and yearly not only wastes the investments made, but also deprives data seekers of the ability to easily access existing data.

- Integration of a comprehensive GIS as an online system that is integrated to the database system would offer a powerful tool in the hands of planners, researchers, and policymakers. The system can further serve in the evaluation phase of plans and policies. Visual presentations of data are always easier to interpret and understand.
- Adopting telemetry systems, which represent the latest technologies in the field of data collection and transmission, can be a major step to improving the quality of data collected from remote locations. It will fill the gaps in data and make data available in the system on time and online. Such a move should take place along with strengthening the national capacity for running and managing these systems.
- Seek and strengthen cooperation with the World Meteorological Organization (WMO), in order to connect with their systems and interact with their new findings on weather conditions, storms, forecasts, and data collection, among other topics of joint interest.
- Strengthening the capacity of human resources is a key issue for the implementation of the improvements required.

Summary

Data availability is essential to formulate and implement water policy, and data collection from implementation provides essential feedback for policy reformulation. Water data systems have acquired increasing importance in water resource assessment, development, management, and update. Water data management iterative systems have facilitated their employment in water policy formulation and follow-up. Jordan recently adopted and is implementing a policy of automated data management systems. Attempts to centralize water data management in Jordan commenced in the early 1980s and gained momentum thereafter. By 2006, water data management systems included reliance on geographical information systems and included laboratory information systems.

Jordan's water strategy adopted in 1997 stipulated the establishment of a comprehensive water databank, along with a decision support system, a monitoring program, and a system of data collection, entry, updating, processing, and dissemination of information. This databank will be designed to become a terminal in a regional databank setup. By 2002, this policy on data management had been substantially implemented, and online telemetry systems were installed.

Data on surface and groundwater resources are collected either manually or automatically on site, transported, and processed at the ministry, but the verification of such data has been lagging behind. Changes in the administrative setup in the water sector adversely affected the data collection efforts because of a lack of funds and different spending priorities. There still exist shortcomings in data management regarding a lack of standard operating procedures and data dissemination.

Measures to improve data quality have been legislated and implemented, a step that would reduce the dependence on estimates and synthesis. SCADA systems have been installed in the Jordan Valley, and general use in the rest of the kingdom is anticipated.

Policy on data publication and dissemination has not progressed as the data systems have. This aspect of water policy is lacking, perhaps because of a combination of reasons. Recommendations to enhance the versatility of the data management system in Jordan are made.

References

Grawitz, B., and Y. Hassan. No date. Water Management Information System (WMIS) for the Jordan Valley. http://www.inbo-news.org/ag2000/jordan.htm (accessed October 6, 2005).

MWI (Ministry of Water and Irrigation). 1997. *Jordan's Water Strategy.* Article no. 10. Amman, Jordan: MWI.

———. 2002. *Ministry of Water and Irrigation Annual Report 2002.* Amman, Jordan: MWI.

———. 2003. *The National Water Master Plan Brochure, February 2003.* Amman, Jordan: MWI.

———. 2004. *Water Resources in Brief.* Amman, Jordan:MWI

———. 2005. *Water Management Information System.* Amman, Jordan: MWI.

Notes

1. Some measurements of surface flows of the Jordan River and other surface streams were made during the British Mandate (1921–1946).

2. The division chief, Salah Darghouth, provided valuable comments.

3. The minister was Dr. Munther J. Haddadin, who drafted the strategy document and debated it with the other parties.

4. The telemetry water quality monitoring pilot project was launched under a bilateral cooperation program with Japan. The Japan International Cooperation Agency (JICA) provided funds and technical know-how to work with Jordan. The Regional Pilot Project for Time Critical Hydrological Data Acquisition and Transmission was financed by the French Global Environment Facility under projects known as EXACT, proposed by the Water Resources Working Group of the multilateral conference of the Middle East peace process.

5. For municipal domestic water, the specifications are stipulated in the Jordanian Standard (JS) 286 of 2001; for industrial waste water, JS 202 of 2004; and for reclaimed domestic wastewater, JS 893 of 2002.

6. For more details about MWI labs, visit http://www.mwi.gov.jo/main%20topics/Labs/Labs/labs.htm.

7. Bylaw number 85 of 2002 (Underground Water Control), issued in pursuance of articles 6 and 32 of the Water Authority law number 18 of 1988. The bylaw was issued in the *Official Gazette* no. 4565, October 1, 2002. Amended through bylaw number 76 of 2003, issued in the *Official Gazette* no. 4608, July 1, 2003. The predecessor of said bylaw is bylaw number 26 of 1977, issued in pursuance to the Natural Resources Authority law number 12 of 1968 and effective until 2002.

8. The system and personnel training were the fruits of a component of bilateral cooperation with the Federal Republic of Germany under bilateral grant assistance administered by the Kreditanstalt für Wiederaufbau (KFW).

9. I was the expert working for UNDP with the Water Authority branch of the Ministry of Water and Irrigation and was instrumental in developing the Water Data Bank of Jordan.

10. USAID's Water Quality Improvement and Conservation Project.

11. These tools were an outcome of a joint project between the Ministry of Water and Irrigation and the Gesellschaft für Technische Zusammenarbeit (GTZ) of the Federal Republic of Germany.

12. A special department in the ministry has been empowered to issue publications releasing data and information important to interested parties. This is done in collaboration with the Water Information System, JVA, and WAJ. The first publications are expected soon.

4

Development of Water Resources and Irrigation

Munther J. Haddadin, Sami J. Sunna',
and Hani A. Al Rashid

The supply of food and water for a rapidly increasing population has been a major concern of the Jordanian authorities ever since the proclamation of the kingdom in 1946. The arid and semiarid climate of the country, along with the limitation on cultivable land and other natural resources, exacerbated these concerns. The 11-year hydrologic cycle is characterized by a 3-year norm above the average and 4 years below it. Therefore, dry spells are inevitable, and measures must be taken to cope with them.

Municipal Water Supply

Traditionally over the years of history, domestic water needs of Jordanians were met from spring flow or by constructing cisterns and ponds to collect and store rainwater. Precipitation supplied soil water needed to support rain-fed agriculture. The growth of villages into towns and cities witnessed the first attempts to build water supply networks, with authority for such projects vested in the municipal councils. Amman, Salt, Madaba, and Irbid are examples of early urban systems of water supply. The health departments in the governorates, branches of the Ministry of Health of the central government, shouldered the main responsibility of assuring that the supplied water was of drinkable standards.

Chapter 2 provides details of the historical development of water institutions and their respective responsibilities. Central government supplied capital assistance or undertook to be the guarantor of loans to municipal institutions to build water projects. When demand for services mounted under population pressures, especially in the case of Amman, attention was focused on the water and wastewater services of the capital city.

Population pressure on water resources countrywide started to show in the early 1960s, but it became visible on municipal water resources in the first half of the 1970s. Water transport from nearby aquifers commenced shortly after the local wells inside and around Amman fell short of meeting the growing demand. Wells were drilled in the Qastal area, about 30 km south of the capital, and water was transported to it. Central government assisted in improving the supply of municipal water through the Domestic Water Supply Corporation. In the Jordan Valley, the Jordan Valley Commission (JVC) and its successor, the Jordan Valley Authority (JVA), were in charge of the municipal water supply (Chapter 2).

The fragmentation of responsibilities exacerbated the challenge of providing a secure and sustainable municipal water service. One important result of fragmentation was that no comprehensive picture of the demand–supply potentials for municipal and industrial water countrywide existed until 1980, when a study of the municipal and industrial needs was concluded. Each of the above authorities went about its tasks separately from the others, and conflicting plans emerged.

First Major Transfer of Municipal Water

Municipal water was given top priority, and supply management was the option in the early years of independence. Demand management measures surfaced only three decades later. When local sources were exhausted, an incremental supply had to come from remote areas, and groundwater was the preferred source for municipal water. This was the case when the city of Irbid, the second largest city in the early 1960s, faced a water shortage. A water transfer project was implemented in 1963 to supply that city with groundwater from the Dhuleil-Azraq aquifer using the abandoned oil pipeline of the Iraq Petroleum Company (IPC).

Because supply management dominated the scene, more supplies became necessary in the early 1970s, and those came from the Aqib aquifer in the northeastern desert. More transfers would come in 1986 from the Jordan Valley, as outlined below.

Competition in Water Allocation

Conditions of shortage triggered competition among water users, and water allocation became an important issue demanding the attention of the central government to arbitrate and decide. A temporary mitigating factor, however, has been mostly financial, because the costs of municipal water transfer projects were beyond the ability of the country to finance from indigenous sources.

Competition between Municipal and Irrigation Uses

Shortages in municipal water supply, accorded top priority by the government, became frequent in urban areas, and per capita consumption started to decline. The government's policy had been to develop agriculture for economic and social reasons. Agriculture had the virtues of employment generation for Jordanians, stabilizing the rural population, and contributing to the economy with a high value-added component. Agriculture benefited from spring flows unused for domestic purposes, with credits extended by the government to encourage farmers to invest in rain-fed and irrigated agriculture.

As Amman began to grow quickly in the 1950s, municipal water started to erode the flows allocated to agriculture. The local spring on which the city depended for its water supply, Ras El Ein, contributed to the base flow of the Zarqa River, the second-largest surface watercourse for Jordan after the Yarmouk. The spring soon dried up as its flow was progressively diverted to municipal use in Amman. Its impact on irrigation was limited to lands downstream of Amman and the Zarqa Gorge. The construction of the King Talal Dam on the Zarqa River helped alleviate the adverse impact on the lower gorge and on agriculture in the Jordan Valley. The irrigated areas downstream of Amman dried up, but those lands were incorporated in the urban zoning, which appreciated their price. However, the environmental impact on the Zarqa Gorge was noticeable.

Water Transfers to Amman

The first instance of official competition over surface water occurred in 1976, while the King Talal Dam was under construction. The Amman Water and Sewerage Authority (AWSA), with approval of the president of the JVA and arbitration by the king, embarked on a project to transfer 10 mcm of surface water from the dam to Amman. The project was shelved, however, because of the potential of environmental hazards.[1] It was replaced at the initiative of the JVA by a larger project to transfer some 45 mcm of Jordan Valley water from the King Abdallah Canal at Deir Alla to Amman. Again, this project was discussed in a large meeting of the highest officials in the country, chaired by the king.

The plan was to divert this flow of surface water from the valley to Amman, where it would be used and then its wastewater collected, treated, and allowed to flow by gravity to the King Talal Dam for reuse in Jordan Valley irrigation. The high-level officials gave their consent. When the project was built by the JVA in 1987, it became the first time that surface water was used for municipal purposes. The design of the project and its operation required foreign assistance not only in financing, which readily came from USAID, the Kuwait Fund, and the Saudi Fund, but also in training of operators.

Competition with the Environment

In the same high-level meeting that approved the Deir Alla–Amman project, and at the suggestion of the JVA, it was decided that a maximum of 12 mcm per year be transferred to Amman from the Azraq aquifer, about 72 km away. With this rate of pumping added to the existing legal and illegal uses of the aquifer, the rate of abstraction approached double the sustainable yield of the aquifer, resulting in lowering of the water table of the Azraq Oasis, a Ramsar site in Jordan known for its value for migratory birds as a desert oasis. Adverse impacts on the environment are not yet fully mitigated.

This undertaking was preceded by other groundwater transfers to Amman from closer areas, tapping Qastal and Swaqa aquifers for incremental supplies to Amman and diverting springs in Wadi Shueib to Salt, and it was later succeeded by more transfers of groundwater, from Wala-Hidan to Madaba and Amman. By 1988, it seemed as though municipal water was denying agriculture and the environment their equitable shares of water. The environmental impact of diverting groundwater from Wala, and drying up the Hidan spring as a result, had distinct negative effects on agriculture and biodiversity in the Hidan Gorge, a tributary of the Mujib. These waters were in the project pipeline of the JVA to irrigate more lands in the Southern Ghors south of the Dead Sea and make more industrial water available to meet the expansion needs of the Arab Potash Company (APC).

More water was diverted from an aquifer in Wadi Arab near North Shuna in the northern Jordan Valley to Irbid, and away from existing agricultural uses.

With water transported over distances and elevations to urban areas, it looked as if the undertaking was an open invitation to rural people in the source areas to follow the water and move to urban areas.

Municipal and Industrial Water Supply and Demand

In 1977, the National Planning Council (NPC), the predecessor of the Ministry of Planning, decided to conduct a study to assess the future situation of the supply–demand pattern of municipal water in the northwest quadrant of the country. The NPC, although not a water agency, had been responsible for international cooperation. In that capacity, it would field questions from donors who supported the water sector. No single government agency was in charge of the water sector, so donors would turn to the NPC for answers.

The study covered the population centers between Wadi Wala in the south and the Yarmouk River in the north, home to about 91% of the population of the kingdom. The results of the study were alarming, even to water officials (Howard Humphrey and Partners 1978). It showed that the needs of northwest Jordan by the year 2000 would be about 250 mcm more than the amounts already supplied from the strained groundwater resources. This conclusion prompted the adoption of a two-pronged policy to secure municipal water supplies until the turn of the century:

1. Allocate water from the Jordanian share in the Yarmouk for municipal supplies in the annual amount of 90 mcm. This could be done only after a dam was built on the river at Maqarin, and it still would not cover the anticipated deficit of 250 mcm.
2. Approach the state of Iraq to Jordan's east for an annual amount for Jordan's use from the Iraqi share in the Euphrates River. Iraq responded positively and agreed to allocate the balance of the deficit, or 160 mcm, in the year 2000.

The above study and the emerging policy showed the importance to Jordan of the construction of the Maqarin Dam on the Yarmouk River, which suddenly became the source of supply for domestic water to north Jordan, including Amman. This fact required that Jordan maintain good relations with Syria, whose approval was needed to build the dam. It also led the Jordanian government to ask the United States to intervene with Israel and obtain its approval to have a diversion structure built across the Yarmouk, downstream at Adassiyya and associated with the Maqarin Dam, to divert water released from the future dam to Jordan. The second prong of the water policy prompted the Jordanian government to cement ties with the government of Iraq to secure the supply of 160 mcm of water, the envisaged shortage by the year 2000, from the Euphrates.

Iraq responded positively, and a feasibility study of a transfer from the Euphrates showed that the water would cost, in 1984 U.S. dollars, about $2.00 equivalent per cubic meter at the outskirts of Amman. This was a higher price than the Jordanians could afford, especially when the cost of distribution was added to it. The project was therefore shelved, and other alternatives were sought.

Water shortage thus promoted regional cooperation and did not trigger conflicts, but it hardly induced cooperation between the government agencies in charge of water, the Water Authority of Jordan (WAJ) and the JVA. Disagreements on water allocation were frequent, and they were resolved by high-level meetings of the state officials chaired by the king.

Both parties had good arguments. The WAJ contended that Amman and environs needed more water, as the city was growing fast; the JVA pointed out that the Southern Ghors region had not benefited from economic and social development, a factor that prompted its inhabitants to migrate to urban areas, primarily Zarqa and environs. A large pocket of poverty had been forming there, and the JVA intended to expand integrated social and economic development in the Southern Ghors after it had implemented the first stage using local surface flows from side wadis to irrigate 4,850 ha.

Disi–Aqaba Transfer

Encouraged by the transfer from Azraq to Amman, the government solved the escalating demand for municipal and industrial water in Aqaba on the Red Sea by transfers from the Disi aquifer. A project was built to transfer 20 mcm/y of freshwater reserves from the Disi aquifer to Aqaba in the 1980s.

The Mujib–Zara–Zarqa Ma'een–Hisban Project

The issue of allocating more water to municipal uses in Amman surfaced again in 1997, when JVA resurrected the Southern Ghors Irrigation Project using the flows of Wadi Mujib, including Wadi Wala-Hidan, and Wadi Hasa floods. The base flow of Wadi Hasa had been allocated and used in irrigation of the first phase of the Southern Ghors development and to augment industrial supplies to the Arab Potash Company. The project, in its second phase, would irrigate 6,000 ha in the Southern Ghors and supply 10 mcm of industrial water to APC. The project was amended in 1997 to include a conveyance line northward by the Dead Sea shore to supply another 10 mcm to tourism development that had started and gained momentum there. These allocations were decided by the minister of water and irrigation, who was in charge of the Jordan Rift Valley development.

At the beginning of the project implementation stage, a portion of the Mujib water was reallocated and added to other sources to secure additional municipal supplies to Amman at the expense of irrigation in the Southern Ghors. The transfer of Mujib water northward to supply tourism development was amended to have the carrier enlarged to accommodate more water resources located along its way. These resources are mostly brackish, with some minor freshwater springs. There is also an uncontrolled flowing well of brackish water in Wadi Hisban of the Jordan Valley. All these resources are being collected in a main carrier and delivered to a desalination plant on the southern bank of Wadi Hisban. The desalinated water will be distributed to the hotels in the amount intended for their use (10 mcm), and the rest will be pumped to Amman over a total head bigger than that of the Deir Alla–Amman project totaling 1,350 m. The transfer rate to Amman will be about 55 mcm/y and is due to be completed in late 2006.

More Water to Amman

The municipal water shortages continued unabated by these transfer projects. The peace treaty with Israel added some 60 mcm of water annually to Jordan and restored the sharing formula on the Yarmouk (Chapter 10). These additional flows enabled the full operation of the Deir Alla–Amman project and paved the way to double its capacity to 90 mcm/y. Simultaneously, a diversion structure has been built across the Yarmouk at Adassiyya, making it possible to control the diversion rate.

Thus municipal water to Amman was given priority over expansion of irrigated agriculture in the Jordan Valley. Return flows continued to be a primary source for irrigation in the middle Jordan Valley to maintain water balance.

Attention then turned to a fossil freshwater aquifer in the south of the country. The Disi to Amman project, meant to supply Amman with 100 mcm annually of fossil water from that aquifer, emerged as the official policy of government after the Euphrates option was shelved. As was the case with the Euphrates transfer, the cost of delivering water to Amman from Disi was prohibitive, and there were environmental concerns as well.

Nevertheless, the government embarked on the project, had it studied for feasibility in 1994–1995, and listed it among its water plans. A discovery of fresh water in the same sandstone aquifer near Karak did not persuade the government to pay attention to possible closer areas of supply. The project was floated for tender among prequalified contractors on the basis of build-operate-transfer (BOT), but the lowest bid was high, and the project is being reassessed as of 2006.

Unlike the current government policy, the authors are of the opinion that the sandstone aquifer south of Amman should be better investigated to determine freshwater occurrence and potential rate of abstraction. There has been evidence that freshwater exists in the sandstone aquifer much closer to Amman than Disi. It may be possible to transfer this water to Amman at less than one-third of the current project cost.

Impact of Water Diversion on Water Resource Development

Jordan's efforts to regulate the Yarmouk River started in 1952 but never materialized for political reasons. The Yarmouk floods continued to flow past the Jordanian diversion tunnel intake toward the Jordan River. Israel was partially benefiting from these flood flows by pumping between 75 and 90 mcm/y to store in Lake Tiberias for later use, much more than its share of 25 mcm.

Under conditions of scarcity, JVA continued to look for alternative solutions to Yarmouk regulation. A reservoir site with the capacity of 55 mcm was identified in the Jordan Valley floor, and an off-river dam, the Karama Dam, was built to divert the Yarmouk floods to the main irrigation canal, the King Abdallah Canal, and discharge them into the reservoir. The project was completed in November 1997, and by April 1998, the reservoir was filled with 87% of its capacity.

The Karama Dam was located on Wadi Mallaha, downstream of the intake of the municipal water to Amman at Deir Alla. The diverted water received treatment in a plant halfway to Amman, operated regularly by trained staff. In July 1998, the raw water had a heavy algae load, and the treatment plant operators could not remove the odor it created. Additionally, microscopic free-living nematodes passed through the sand filters. These were duly disinfected with chlorine, and nothing in the finished water specifications prohibited the existence of dead free-living nematodes. The odor came under control four days later, but the dead nematodes in the finished water caused an uproar in the press motivated by political considerations. The political parties opposed to peace with Israel saw it as an opportune moment to blame the event on water coming from Israel, even claiming it was wastewater. The traditional politicians in the country saw it as an opportune moment to attack the government in the hope of bringing it down and enhancing their chances of forming another government instead. The king was at the Mayo Clinic in the United States for medical treatment, denying the government a fair arbiter. The regent took the heat and became a target. The minister of water and irrigation resigned, and nine days later the government fell. Several water officials and one health official were charged with negligence by the succeeding government, but all of them were cleared three years later.

This event, unprecedented in Jordan's history, spread fear among water professionals and affected the spirit and performance of them all. It also had adverse consequences to the Karama Dam. Floods containing high silt contents were disposed of before they reached the municipal water intake at Deir Alla for fear of excessive nematode counts. This uneducated move denied the Karama Dam the floods it was designed to impound, and the reservoir remained empty for about three years. The situation was finally mitigated by a better assessment and better judgment by water officials. The floods were allowed to pass in 2003, and the reservoir was full during the 2003–2004 season.

On the other hand, the environmental requirements in the lower Wadi Mujib led to an increase of the cost per cubic meter of water for all its allocated quotas: irrigation, industry, and municipal supplies. The existence of rare species of aquatic life required that the stream flow of the wadi be maintained to preserve these species. Consequently, the upstream structure that would have allowed the diversion of water by gravity to its destinations was eliminated, and the stream flow has been allowed to continue to a point short of the Dead Sea shore, where water diversion will be done by pumping. This issue is discussed further below as the competition of the environment for water is addressed.

Industrial Competition for Water

The competition for water with irrigation uses was not limited to municipal uses; in practically all cities of Jordan, industries also got their water from the municipal network. The JVA faced yet another challenge in 1986, posed by the Arab Potash Company, which was operating in the Southern Ghors. The APC's Dead Sea Arab Potash Project faced a shortage of water for its ongoing production as a result of declines in groundwater levels in wells supplying the company. The JVA made this an important priority for economic, social, and political reasons. This project provided jobs for the locals; had been a foreign currency earner, as it exported all of its products; and was owned by several Arab countries in partnership with Jordan. Irrigation water was so managed as to transfer decent amounts to the APC, especially during off-peak irrigation demand. This arrangement helped meet the company's needs until it embarked on expansion of its production in 1988.

The APC then called on domestic expertise to help solve its water supply problem. Tests were conducted on the use of a modified industrial process, cold crystallization, to refine brackish water and produce potassium chloride. This worked, and brackish water was readily available in the proximity of the company's factory in the Southern Ghors. However, more freshwater resources were needed to expand production of magnesia, bromine, table salt, and other minerals. Those needs were accounted for in the Mujib Project, which was completed in early 2005, and the industrial water needs have been supplied to the APC ever since.

As a result of competition with industrial and municipal purposes, irrigation was left with less than half the intended water for development. However, the high cost of the Mujib Project and its feasibility analysis dictated that other paying customers receive sizable portions of the water to justify it. This fact mitigated the competition among the different users of the project's water.

Environmental Competition for Water

Municipal uses competed with agriculture and the environment, as described for the Wala–Amman municipal water project. In the Southern Ghors Irrigation Project, resurrected in 1997, three dams were to be built. One, on Wadi Wala, was meant to recharge the groundwater aquifer and revive the Hidan Springs. By 2004, the spring flow emerged as a result of groundwater recharge, the base flow of the Wadi Hidan returned, and color started to come back to the gorge, showing that the environment was recovering.

The Mujib Project called for diversion of its regulated flow by a weir situated shortly after Wadi Hidan meets Mujib and a few kilometers before the Mujib discharges into the Dead Sea. The JVA proceeded with the implementation steps, with funding provided by the Arab Fund for Economic and Social Development and the Treasury of Jordan, after which environmental concerns were voiced over that part of the project. Before the creation in 2001 of the Ministry of Environment, environmental issues were strongly advocated by a nongovernmental organization supported by the king called the Royal Society for the Conservation of Nature (RSCN), established in 1965. The RSCN contended that drying up the Wadi Mujib below the diversion structure would destroy aquatic life, some of which was unique in the world. This meant that Mujib water should be allowed to flow in its gorge until it came close to its terminus in the Dead Sea.

Project economics were tight, although the added capital cost was not as much of a problem as was the added operational cost. Water had to be pumped to serve the users in the Southern Ghors and the hotels up north, a high energy cost that would endanger the feasibility of the project on the one hand, and impose higher water charges on farmers who would not be able to meet them on the other. The RSCN mobilized support from various influential parties, including the World Bank. They all supported the environmental concerns, and the project was modified to divert the water at a point just before the Wadi discharges into the Dead Sea, expand it to collect the brackish sources on the way, desalinate all of the water, and pump it to Amman. A challenge remains in project cost recovery, however, although the modification brought in Amman consumers as beneficiaries, which would improve the ability of beneficiaries to pay.

Irrigation Water Supply

Historically and for millennia, the Jordan Valley had been a region suited for irrigated agriculture, with unique climatic properties that enabled the production of winter fruits and vegetables. Irrigation water was drawn from the side wadis discharging into the Jordan River, from the river itself, and from its major tributary, the Yarmouk. Archaeological remains provide evidence of early farming society in the Jordan Valley. Communities existed in the valley since around 10,000 BC (Khouri 1988).[2] A decline in irrigated agriculture took place for a variety of reasons, including wars, epidemics (especially malaria), drainage problems, and droughts.

In modern times prior to the 1960s, rain-fed agriculture was the backbone of food production in Jordan. The relatively low intensity of population in the past few centuries had created a balance in the supply–demand equation for food commodities, with surpluses of grain exported to Palestine and fruits imported from there. Irrigated farming was resurrected in the early 1950s, and the networks were mostly surface earth canals, with a few concrete-lined canals where government was involved, as in the Deir Alla agricultural project of 1954. On-farm irrigation methods were primitive, and farmers used furrow and basin irrigation methods and open fields.

The Deir Alla Irrigation Project

Perhaps the first involvement of central government in an irrigation project was by the Department of Irrigation in 1954. It was the Deir Alla Irrigation Project in the Jordan Valley, irrigating some 500 ha with water from the Zarqa River. An agricultural experiment station was established there and became active in research and extension inasmuch as the capacity of the country allowed. Expatriate experts were recruited by the government to help design and manage irrigation projects in addition to a limited number of nationals.[3] The policy adopted at that time was to have the central government invest in conveyance and distribution systems and leave the responsibility of water application on the farm, along with the farming duties, to the farmers. The central government, under this policy, would involve itself in the provision of the services of agricultural research and extension.

The East Ghor Canal Project

The first major undertaking by the government in building irrigation infrastructure was the comprehensive study of the Yarmouk–Jordan Valley Project, for which it commissioned a joint venture of two American firms (Baker and Harza 1955). A master plan was prepared in 1955 and served as the basis for the Jordan Valley irrigation development.

Through the late 1940s and 1950s, Jordan experienced surges in population levels (Chapter 1). Irrigated agriculture in the valley helped the government achieve its objectives of creating jobs for an expanding population, using as much as it could of its water rights, populating the border area with Israel, and increasing agricultural production. Grants from the United States between 1959 and 1966 made possible the irrigation of 11,400 ha at a total cost of US$12 million. The free gravity flow of a lateral drop inlet on the southern bank of the Yarmouk, close to the Jordanian village of Adassiyya, diverted water into a 1-km-long tunnel that saw daylight at the beginning of the East Ghor Canal.

Extension of the irrigation network continued in 1968, and an additional 1,100 ha were included in the surface irrigation network of concrete-lined canals. The main canal length reached 70 km and 14,000 ha by 1969, 78 km and 22,300 ha by 1978, and 110.5 km and 28,300 ha by 1987.

The farms used furrow and basin irrigation methods under gravity conditions. Off-season fruits and vegetables became available on the local market, with surpluses for export, primarily to Syria and Lebanon, and later to the Gulf states.

A New Rural Development Approach

It took four years after the June war of 1967 to restore law and order in the country. In early 1971, the government, upon directives from the king and his crown prince, initiated efforts to plan for the rehabilitation and development of the Jordan Valley–East Bank. A new strategy was adopted by which water would play the major role. The development of the Jordan Valley would be undertaken in an integrated social and economic fashion for the first time. The strategy aimed to rehabilitate the Jordan Valley in the wake of the destruction it had suffered during the turbulent years after the war, as well as to further develop it, increase agricultural production, create more job opportunities in various sectors of the economy, bring back the valley's inhabitants, and attract other citizens to take up residence and work there. The population would have a much better standard of living and more than one reason to stay in the valley, which Israel desired to depopulate and, if practical, to occupy. A populated valley would be harder for an occupying power to swallow.[4]

The Jordan Valley development effort was expanded in 1977 to include the Southern Ghors south of the Dead Sea. The same strategy was applied, and the region witnessed, for the first time in its modern history, the benefits of development. The irrigated area became 4,900 ha by 1986.

The backbone of this effort was to be irrigated agriculture. This policy was implemented through successive development plans laid out by the government institution in charge, the JVC and its successor, the JVA. The valley reached its "cruising altitude" in 1987. By that year, a total of about 200 mcm/y on average was allocated and used in irrigated agriculture in the Jordan Valley, including the Southern Ghors.

A total of nine dams were built in the process of this development, with a total capacity of about 170 mcm. A major dam is being built on the Yarmouk, the Wehda Dam at Maqarin, with a planned yield of 120 mcm. By the completion of this dam, all surface water resources will have been regulated.

The economic, social, cultural, and political impacts of the integrated rural development are presented in Chapter 9.

Irrigation on the Highlands

The success of irrigated agriculture in the valley prompted landowners outside the valley to obtain permits to drill for groundwater and use it in irrigation. The government policy was clear in supporting agricultural production and the agricultural sector at large. The strategic importance of agriculture, the government considered, went beyond the economic gains. It had immense social, demographic, and environmental gains that could outweigh the economic benefit. Some owners violated the law and had wells drilled without permits. Such wells were scattered over the kingdom and have caused problems to managers of groundwater. The policy on groundwater management was clear, but the implementation mechanism was not.

By 1984, fossil groundwater was abstracted for use in irrigation. The Disi-Mudawwara aquifer was open to agricultural investments that year (Chapter 1), and wheat fields covered 8,000 ha by 1989. The cropping pattern was amended to include some stone fruits and vegetables. Water consumption in 1992 was estimated at about 85 mcm/y. Issues regarding the need to diminish those flows were outlined in Chapter 1.

By 1987, the irrigated area outside the Jordan Valley had matched and surpassed the irrigated area in the valley. Luckily, the production seasons in the valley did not coincide with the seasons outside it, with the exception of some vegetable production from the Disi area. Fruits and vegetables became available as fresh products in the kingdom between December and October of each year, with the Jordan Valley contributing off-season crops from December to June, and the Jordanian Highlands and Plateau contributing from June to October. Excesses in the production of tomatoes prompted the government to import and erect two tomato paste factories for the Jordan Valley in 1982.

Balancing Municipal, Industrial, and Agricultural Water

The expansion in water needs for all purposes paralleled the increase of population by both natural growth and waves of refugees. Although the natural renewable water resources are limited and fluctuate in availability around a certain annual average, the population numbers, standard of living, and available technologies have been on the rise, making the balancing act among competing demands all the more difficult.

Jordan experienced still more factors beyond its control that further exacerbated the imbalance in the population–water resources relationship. Its shares in the international watercourses, the Yarmouk and the Jordan, and transboundary groundwater were subject to increasing use by other riparian parties, to the disadvantage of Jordan. The available freshwater supplies were diminishing, while the demands were rising.

The government policy of interbasin water transfers was coupled with a more daring policy aiming at balancing the demands for municipal and irrigation water. The reuse of treated municipal wastewater emerged as a government policy as outlined above. It compensated the Jordan Valley with about 65% of the water that was pumped to Amman. More treated wastewater was flowing to the Jordan Valley than the effluent of its own contribution to Amman's municipal water. Amman's consumption of municipal water produced effluent that discharged naturally into the Zarqa River Basin and the King Talal Dam reservoir. Treated effluent elsewhere, such as Irbid, Mafraq, Madaba, and Karak, was put to use in local irrigation by private farmers.

Simultaneous with the above balancing of allocations to irrigation and municipal uses, the country's shares in international water, primarily in the Yarmouk, were being eroded by acts of Syria and Israel. The government responded in a pacifistic fashion, avoiding clashes with the neighbors over water sharing. It pursued diplomatic means with Syria, to little or no avail, and managed the conflict with Israel through sponsorship of the United Nations Truce Supervision Organization (UNTSO), a mechanism that yielded positive results most of the time.

Resource Management under Stress

The water policy of the government shifted to crisis management, in which irrigation water was rationed with a tilt in favor of perennial crops. However, the backbone of the water policy stressed the approach of demand management through improvement of water use efficiency and management structure.

Municipal Water Management

Since the early 1970s, the automatic response to shortages in municipal supplies has been rationing of water service. The next option, prompted also by the need to recover the cost of service, has been demand management, adjusting the water tariff upward with due consideration of affordability to the poor. All connections to the network in Jordan are metered.

Low water efficiency often stems from water losses caused by leakage from old networks, illegal connections, and errors in metering. Improving efficiency requires competent human resources; financial outlays for operating, maintenance, and replacement costs; and improved technology in network surveillance, reporting of breakages, and response to emergencies. Attention has been paid to all these factors to improve efficiency. The water tariff was adjusted periodically. A center was established for the training of WAJ operators, vocational training is contributing to the supply of technicians, and managers are sent overseas to participate in seminars and courses to upgrade their abilities. The old networks of major cities are being replaced, with pipe material selected to make it difficult to install illegal connections. Metering is being improved through the introduction of modern technology.

In Amman and environs, the water and wastewater administration has been outsourced (Chapter 2) to improve management and response to emergencies. Maintenance facilities have been upgraded. The current trend in water management is to outsource management and, where possible, form private companies to undertake the municipal water and wastewater service, as has been done in Aqaba.

Irrigation Water Management

The JVC, upon urging by World Bank staff on their field mission, decided in the summer of 1973 to replace a plan to extend the East Ghor Canal by 18 km to irrigate 3,600 ha by a surface distribution network with another plan. The measure signaled a change in policy regarding irrigation infrastructure. The policy sought improvement of distribution efficiency by adopting pressure pipe networks. The donor, USAID, concurred and encouraged the shift (JVC 1973). The design of the new network made use of the differential head between the water source (a side wadi or the East Ghor Canal) and the farm level to generate free energy, supplementing the head where needed with booster pumps.

Farmers in the valley, especially in the newly irrigated areas, responded by adopting drip irrigation as their preferred on-farm system. The Agricultural Credit Corporation, a government agency, encouraged the trend and provided loans to farmers to procure the advanced systems. The introduction of plastic tunnels is credited to the Deir Alla Agricultural Experiment Station in the early 1970s. Soon factories to make drip hoses were set up in Jordan, and the adoption of drip methods became widespread. Plastic houses were introduced in the process, and the productivity per unit land area and unit water flow increased substantially.

The use of pressure pipe systems for the conveyance of irrigation water was then generalized. Projects to rehabilitate the old open-canal systems in the valley by replacing them with pressure pipe networks were prepared and implemented by the JVA between 1986 and 1996.[5] As a result of this policy, all government-sponsored irrigation projects consisted of pressure pipe network. The returns from horticultural crops and the water shortage prompted farmers to revert to drip systems. The combination of pressure pipe networks and drip systems, along with microsprinkler systems for trees, boosted the overall irrigation efficiency from 45% to about 75%. This improvement in irrigation efficiency helped offset the reduction in international water shares, assisted in balancing the allocation of water between irrigation and municipal uses, and promoted unprecedented increases in agricultural yields.

Irrigation water rationing has been practiced as well, especially in the dry months. Farmers have to make ends meet or reduce the area of their plantations if all measures to conserve water do not help.

Sustainability of Irrigated Agriculture

Perhaps the highest risk to the sustainability of irrigated agriculture lies in the Jordanian Highlands, where irrigation water is abstracted from groundwater reservoirs. Overabstraction from practically all the aquifers in the kingdom risks the gradual deterioration of groundwater quality. To minimize that risk, the government decreed in 1992 that no more well drilling aiming at the expansion of groundwater abstraction would be permitted. Exceptions were wells needed for educational institutions, primarily universities, and for factories. The ability of the government to enforce this policy was impaired by several factors, however, among them the quick turnover in officials, hesitant law enforcers, and the general mood of the people as a result of turbulence in the region involving Palestine and Iraq.

Another threat to the sustainability of irrigated agriculture is the further reduction in supplies resulting from upstream uses by other riparian parties, primarily Syria, and from drought conditions. Surface water quality could be yet another threat. The government's policy has been to defend the shares in international watercourses and invest in upgrading wastewater treatment plants to improve water quality, especially in the Zarqa River.[6] Policy to enhance sustainability of irrigated agriculture includes the introduction of biological control of pests in lieu of excessive use of pesticides; cooperation between research and extension services, and between them and the farmers and the marketing institutions, in an effort to avoid successive losses by farmers and enhance their net profit; and encouraging land reclamation and leaching of excessive salts in the root zone of crops by installing drainage networks and granting the farmers free fresh water in winter to leach their soils.

Development of Rain-fed Agriculture and Green-Water Management

The rain-fed strip of the country, running from north to south, is part of the lower western end of the Fertile Crescent. The distribution of rainfall shown in Figure 1-1 tells of the parallel distribution of rain-fed farming. The arable lands that fall within the line of 350 mm annual intensity are fit for rain-fed farming, although some crops can grow with 300 mm rain intensity. Outside this intensity, supplementary or virtually total irrigation is needed to have any decent agricultural yield. However, natural rangelands spread outside these areas, thus contributing to the support of livestock.

In irrigated agriculture, blue-water application to the soil is an artificial attempt to create green water (soil water), whereas in rain-fed farming, water application is done by nature through precipitation that penetrates the ground surface and forms green water. The linkage between agriculture, both rain-fed and irrigated, and water is therefore very clear. Irrigation blue-water flow can be easily measured, but in rain-fed agriculture, measurement of green water is complex if a rigorous analysis is pursued. In semiarid regions, productivity of green water by precipitation is low in comparison with that seen in irrigated agriculture, and better control can be exercised with irrigation. The irrigation water equivalent of soil water is not easy to calculate. However, in the case of Jordan and similar lower-middle income countries, some estimates of the irrigation water equivalent in cubic meters per year are made and described in Chapter 7.

Inverse Relationship between Urbanization and Green Water

Jordan has witnessed rapid urbanization, and today about 70% of Jordan's population lives in urban areas. Settled Jordanians have occupied villages close to or on rain-fed agricultural lands since ancient times. Archaeological remains are witness to this and to the rationality of sites occupied by human settlements. Ancient populations tended to build their settlements on nonagricultural lands, leaving the agricultural lands with their green water for cultivation. This was mostly the case in Jordan until the middle of the last century. Then a sudden influx of refugees and displaced persons defied rational urban planning. Refugee camps sprang up in Jordan, and most of them were built next to urban areas on good agricultural lands.

The economic and social development of the country, along with natural population growth, generated a high demand for housing and other buildings and facilities for services. People migrated from rural villages to urban centers. Towns and cities grew rapidly, and most of the expansion was horizontal and at the expense of good agricultural lands. Such was the case with Amman, Salt, Irbid, Sareeh, Husn, Howwarah, Madaba, Qasr, Rabbah, Karak, and other settlements. Not only did the expansion in urbanization swallow agricultural lands, which basically are reservoirs for soil (green) water, but the concrete surfaces and pavements impaired the natural recharge of groundwater reservoirs as a result of surface runoff. In water terms, the acquisition by urbanization of 400 ha of agricultural land translates into the loss of the equivalent of 1 mcm of irrigation water per year, if not more (detailed in Chapter 7). Thus in most urban areas of Jordan, especially those located on Khatt Shabeeb, a clear inverse relationship exists between urban expansion and sustainability of rain-fed agriculture.[7] Urban expansion means loss of agricultural land and the annual green water it stores from precipitation.

It was this inverse relationship that prompted the JVA to plan settlements in the Jordan Valley on nonarable lands. Expansion of existing settlements was possible only if there was room to expand on nonarable lands. Exceptions were very few, and the policy was strictly reinforced until late in the last century, when the responsibility shifted from the JVA to the Ministry of Municipalities. Somehow, law enforcement responsibility got lost in the process, and the spread of unlicensed buildings occurred.

Hydropower Generation

The use of water resources for hydropower generation is limited in Jordan. A hydropower plant exists at the toe of the King Talal Dam, but its operation coincides with the schedule of releases for irrigation. There is no facility for reregulation to allow generation of power at will. The value of the power thus generated is less than optimum. The power plant is connected to the national grid, but the produced power is both modest and untimely.

The Wehda Dam will have a power generation facility, but most of its production will be the share of Syria. It will have the same handicap as the King Talal Dam in the sense that power generation will respond to irrigation releases rather than peak demand. This aspect of partnership between Jordan and Syria in the Wehda Dam may cause conflict between the interests of both parties: Jordan's in irrigation releases and Syria's in power generation.

Treated Wastewater Reuse in Agriculture

Return flows from municipal water use to groundwater aquifers were addressed in Chapter 1. The intentional reuse of treated wastewater in agriculture started in 1968, almost immediately after the installation and operation of the first wastewater treatment plant in the country at Ein Ghazal, on the periphery of Amman at that time. It was a mechanical treatment plant for the city of Amman, and it discharged its effluent into the once beautiful stream of Seil Amman, which had since dried up. Farmers on both sides of the newly recharged stream started to irrigate crops, without government consent, as they had when the freshwater stream was there. In the late 1970s, an outbreak of cholera prompted the government to plow the fields that used treated wastewater for the production of crops eaten raw. In the following seasons, however, the same farmers resumed their practice and got away with it as long as there were no outbreaks of disease.

The government formally endorsed wastewater reuse in 1977. The reduction in the freshwater supply to the Middle Ghors of the valley could be mitigated by the reuse of treated wastewater. The construction of King Talal Dam on the Zarqa River helped in formulating that policy. The dam reservoir was a perfect place for blending the wastewater with the impounded flood water and for regulating the flow of the Zarqa River. Furthermore, the adoption of piped distribution networks for the region's irrigation system and of drip irrigation techniques by farmers reduced the exposure of farm workers and operators to the blend of flood water and treated wastewater.

Competition over Treated Wastewater

Treated wastewater thus became an irrigation resource in the water budget of the Jordan Valley, and the JVA counted on it as much as freshwater sources such as the Yarmouk River. In the early 1980s, after an extended drought period, the treated wastewater became attractive to the officials of the WAJ, which owned and operated the wastewater treatment plant at Khirbit As-Samra. The WAJ wanted to use part of the effluent to irrigate trees around the treatment plant and proposed to use the effluent to irrigate fields on the plateau close to the plant instead of letting it flow to the Jordan Valley. Such plans would use water at the expense of the Jordan Valley's irrigation water budget.

The dispute in 1983 between the presidents of the JVA and the WAJ had to be settled by the prime minister, who was amused that the two officials were having a clash over wastewater.[8] He ruled that the effluent was part of the Jordan Valley's water resources.

Expansion in Wastewater Reuse

The construction of wastewater treatment plants expanded to include other cities, such as Aqaba, Salt, Irbid, Ajloun–Kufranja–Ein Janna–Anjara, Madaba, Karak, Nuaymeh–Kitm–Shatanah, and Wadi Musa. Zarqa and environs were connected to the Khirbit As-Samra treatment plant, to be added to Amman influent. The Amman–Zarqa area required an expansion in the treatment facilities, and another wastewater treatment plant is being built to accommodate the hike in influent flow.[9]

The technology used for wastewater treatment was the activated sludge technique at Salt and at Amman's first plant at Ein Ghazal, followed, in the case of Amman, by stabilization ponds employing solar energy. The stabilization pond technique was used at Khirbit As-Samra, which has been overloaded since 1990.[10]

It is fortunate for Jordan that its urban areas are located in the catchment of the Jordan River. The wastewater effluent flows toward the Jordan Valley by gravity, without the need for pumping. When the water is diverted at appropriate elevations into the irrigation conveyance systems, the hydraulics of the networks would allow pressures of up to 3 atm on certain farms at lower elevations. The head of 3 atm was the design pressure on the farm level. The gravity head is used to the fullest and is supplemented with pumping where necessary to develop the design pressure on farms closer to the diversion weirs. In the case of Aqaba, the WAJ has implemented, in cooperation with the University of Jordan, an irrigation project next to the wastewater treatment plant where palm trees were planted.

The annual flow of treated wastewater output in 2002 was 80.62 mcm, of which about 73.70 mcm were used in agriculture. Water yet unused amounts to 6.92 mcm.

Future prospects for wastewater reuse are wide. The Ministry of Water and Irrigation (MWI) has made some projections for wastewater reuse in the coming two decades, as shown in Table 4-1.

The water resource policy requiring the treatment and reuse of municipal wastewater is compatible with the rational use and reuse of water resources. Arid and semiarid countries cannot afford to treat wastewater as waste. It should be treated to standards allowing its reuse in irrigation and other purposes.

Summary

The fragmentation of responsibilities for municipal water supply exacerbated the challenges faced in securing water sources. When local and nearby sources ran out, water was transported to urban areas from afar, such as in the case of

TABLE 4-1. Projected Flows of Wastewater (mcm)

	2005	2010	2015	2020
Inflow to treatment plants	146	181	215	247
Treatment effluent	138	170	202	231
Effluent inflow into reservoirs[a]	−71	−86	−100	−114
Total wastewater contribution[b]	67	84	102	117

a. Counted as part of base flow (surface water) and reservoir yields.
b. Increment in water resources over and above reservoir yields.
Source: MWI 2004.

Irbid, since the 1960s, and later Amman and Zarqa, since the late 1970s. Competition between agricultural and municipal water uses worsened as the imbalance in the population–water resources equation reflected an increasing deficit. Diversion of agricultural water to municipal uses was first practiced in the capital city of Amman in the 1950s. Municipal water to Amman was accorded highest priority, and projects were implemented to transfer water to it from nearby aquifers, Azraq, and then the Jordan Valley. Almost simultaneously, similar projects were implemented for Irbid from groundwater sources in the Jordan Valley. Aqaba received transfers from the Disi aquifers as well.

The water policy shifted to accommodate municipal and industrial water needs. Planning for agricultural expansion had to give way when allocations from the same sources were needed for municipal and industrial uses. This also prompted Jordan to adopt a policy of regional cooperation over water.

More friction developed over water allocation between officials in charge of the development of the Jordan Valley and officials of the Water Authority of Jordan. The disputes were resolved in favor of allocating resources that had been earmarked for irrigation to municipal water for Amman and Madaba. More unused resources were shared between the user sectors when the Mujib Project was implemented.

Industry also began competing for water in 1984, when the Arab Potash Company needed increased amounts. Coordination between the Jordan Valley Authority and the industries managed the situation to the advantage of both. The environment also competed with agriculture and municipal uses, particularly in the Mujib Basin, and resolution was made in favor of the environment. The role of donor agencies in conflict resolution was always helpful.

Development of irrigated agriculture commenced on a formal scale in the Jordan Valley in response to competition with other riparian parties, particularly Israel. Irrigation infrastructure was developed by government and advanced from surface irrigation conveyance systems to piped irrigation networks. Parallel advances were made by the farmers. Advanced piped on-farm systems have proliferated since the mid-1970s. The results have been savings in irrigation water use, increased yields, and better-quality products.

Under conditions of scarcity, both supply and demand management measures were adopted. However, the development of irrigated agriculture in Jordan's Highlands progressed unabated as the private sector drew on groundwater resources. Irrigated areas expanded and became almost equal to the areas in the Jordan Valley. Overabstraction from aquifers for agricultural uses threatens their sustainability. Agriculture in the Jordan Valley faces threats to its sustainability from increased Syrian abstractions from the Yarmouk Basin. More attention is needed to manage green water stored annually in cultivable soils.

Under conditions of water stress, competition arose in Jordan over treated wastewater reuse between Jordan Valley agriculture and agriculture on the Highlands. A more pronounced role awaits treated municipal wastewater.

References

Baker, M., Jr., Inc., and Harza Engineering Company. 1955. *Yarmouk Jordan Valley Project: Master Plan Report*. Amman, Jordan: Ministry of Finance, submitted to Jordan–United States Technical Services.

Howard Humphrey and Partners. 1978. *North Jordan Domestic Water Supply*. London, England: Howard Humphrey and Partners. Study conducted for the National Planning Council (Jordan).

JVC (Jordan Valley Commission). 1973. The 18 Kilometers Extension of the East Ghor Canal. Project funded by USAID loan no. 278-H-009.

Khouri, R.S. 1988. *The Antiquities of the Jordan Valley*. Amman, Jordan: Al Kutba Publishers.

MWI (Ministry of Water and Irrigation). 2004. *Water Resources in Brief*. Amman, Jordan: MWI.

Notes

1. Munther J. Haddadin was instrumental in canceling the project because of quality hazards that were looming on the horizon. He was then vice president of the JVA.

2. Khouri documents from archaeological references some 47 ancient settlements in the Jordan River Valley, 16 settlements in the Southern Ghors, and 13 settlements in Wadi Araba.

3. Experts included the British subject Michael Ionides, famed for his Jordan Valley development plan of 1939, and a Belarus native and British national Nicholi Simansky in the early 1950s. Nasoh Al Taher led the nationals, who were, in addition to him, Izz Eddin Yunis, Najib Tleil, and Sweilim Haddad.

4. Israeli troops actually crossed the Jordan on March 21, 1968, and attacked settlements in the Jordan Valley, particularly a refugee camp at Karama. The Israeli troops were pushed back by the end of the day by the Jordanian armed forces.

5. A project replacing the networks in 6,000 ha was financed by the Arab Fund for Economic and Social Development in 1986; another for a similar area was funded by the Overseas Economic Cooperation Fund of Japan in 1992.

6. A wastewater treatment plant contract was awarded in 2004 at a cost of the equivalent of US$150 million to upgrade the quality of effluent of the liquid wastes from Amman and Zarqa to the benefit of irrigation in the Jordan Valley.

7. Khatt Shabeeb is the name given to the north–south agricultural strip that forms part of the western end of the Fertile Crescent. (The rest of that end is in Israel and Palestine.)

8. The president of the JVA told the prime minister that the effluent had become a water resource, and thus wastewater could not be treated as waste any longer. This statement became a government policy since then.

9. The new plant is a mechanical treatment system, built after the build-operate-transfer mode, with about half the cost from a USAID grant and the government Treasury and the other half provided by the contractor. Work started in early 2004.

10. The plant was designed to treat an influent of 68,000 m³/day, but the influent got as high as 170,000 m³/day by the turn of the century. The plant was upgraded then, and the new plant will improve the performance.

5

Environmental Issues
of Water Resources

Ra'ed Daoud, Helena Naber, Mai Abu
Tarbush, Razan Quossous,
Amer Salman, and Emad Karablieh

Jordan is part of the Levant, a term referring to the countries bordering on the eastern Mediterranean. This region has provided biogeographic and trade links among Europe, Asia, and Africa. The natural environment in Jordan is mainly semidesert land, but despite this and the limited area of its territories (89,322 km²), it is home to four distinct, highly diverse biogeographic regions: the Mediterranean Region, the most fertile part of Jordan and home to most of its main cities and towns; the Irano-Turanian Region, which is phytogeographically a narrow strip of variable width that surrounds the Mediterranean ecozone on the east, south, and west; the Badia Region, a biogeographic region consisting of the stretch of the Eastern Desert encompassing almost 75% of Jordan's area; and the Sudanian (or Subtropical or Afrotropical) Region, located in the Jordan Rift Valley between the town of Karama, some 40 km north of the Dead Sea, and the Gulf of Aqaba in the south. This biogeographic classification closely corresponds with the geographic classification presented in Chapter 1, where the country is said to have three distinct geographic regions: the Jordan Rift Valley, the Highlands, and the Badia.

Jordan's Ecosystems

These biogeographic regions with their particular environmental conditions give rise to several unique ecosystems in Jordan (MOE 2003).

Desert Ecosystem

The desert ecosystem covers approximately 75% of Jordan's area, forming the country's main rangeland. It is a gently undulating plateau with an elevation of 500 to 900 m above sea level and is characterized by an arid climate, hot and dry in summer and cold in winter, with a substantial day–night temperature variation. The vegetation cover, located mainly in wadis, is very poor and unproductive. However, on its highlands ecosystem side, it has several Irano-Turanian species of small shrubs and bushes, such as *Artemisia, Retama, Anabsis,* and *Ziziphus.* Four broad habitats may be identified in this ecosystem: the smooth gravel of the Hamad Desert; the Harrah Desert, identified by its black basaltic rocks; sand dunes; and clay pans lying at the bottom of closed drainage basins.

Scarp and Highlands Ecosystem

The scarp and highlands ecosystem is characterized by the mountainous strip that adjoins the Jordan Valley to the east. The climate in this ecosystem is mostly Mediterranean: moderate and dry in summer, cold and humid in winter. The altitude in this area ranges between 600 and 1,500 m above sea level. It contains the largest remaining areas of natural forest, dominated by *Pinus halepensis* at elevations above 700 m and mixed evergreen–deciduous oak forest of *Quercus calliprinos* and *Q. aegilops* at lower elevations.

Subtropical Ecosystem

The subtropical ecosystem stretches through the Jordan Valley, which is part of the Great African Rift. It is the westernmost strip of the country, ranging in width from 10 to 20 km and extending 370 km in length. It has the lowest dry contour on earth, at 418 m below sea level in 2005, marking the level of the Dead Sea. The Jordan Valley is a junction for important biodiversity, acting as a convergence point of the main global biogeographic regions: the Palearctic, Ethiopian, and Oriental.

Dead Sea Basin Ecosystem

The Dead Sea Basin ecosystem is located geographically within the Jordan Rift Valley, situated on the shores of the Dead Sea and at the oases in its vicinity. This ecosystem is an important bird breeding area.

Jordan River Basin Ecosystem

Located in the northern part of the Jordan Valley, the Jordan River Basin ecosystem consists of the Jordan River and its tributaries. The area serves as a major route for migrating birds and is characterized by many endemic and globally valuable species, such as the brown fish owl, common otter, Arabian leopard, rock hyrax, fresh water turtle, and several freshwater fish. It contains the majority of the country's surface-water resources.

Gulf of Aqaba Ecosystem

The Gulf of Aqaba ecosystem is a marine environment, unique because of its tropical coral reef—the northernmost such reef—as well as its sea grasses and fish, many of which are endemic. The Gulf of Aqaba is also an important bird area, as it forms a bottleneck site through which millions of birds pass in their seasonal migration.

Freshwater Ecosystem

The major freshwater ecosystem in Jordan is the Azraq Oasis, located in the Eastern Badia region about 75 km northeast of Amman. It is protected by the Ramsar Convention, entailing commitments under this agreement to conserve wetlands, and is considered an important oasis for indigenous bird life and a station for migratory birds.

Land Use

The Ministry of Municipal Affairs is responsible for land use for urban purposes and oversees the preparation of town plans over most of the country; land use and planning are the responsibilities of the municipality of Amman and the Aqaba Special Economic Zone Authority in those two cities. Land use and planning for agricultural and grazing lands and forests are under the jurisdiction of the Ministry of Agriculture. Coordination among the different agencies responsible for land use in the country is achieved through the formation of multiagency committees and the issuing of permits for various uses.

The land use planning experience in the Jordan Valley represents the kingdom's best example of coordinated land use policy. The responsibility was vested in the Jordan Valley Authority (JVA) for more than a decade and a half between 1973 and 1988, during which the JVA delineated nonarable lands for the preparation of town plans and limited the expansion of existing towns and villages to the extent that the expansion would encroach on arable or cultivable lands. It prepared the town plans for some 45 settlements and supplied them with all amenities and public utilities.

However, in other parts of the country, and because most towns were originally situated next to arable lands, the expansion of cities and towns, even villages, encroached on good agricultural rain-fed lands and natural areas. The loss of such lands entailed the loss of water resources as well. Rain-fed agricultural soils are also reservoirs for soil water retained in them through capillary action and surface tension. The good soil, along with the corresponding soil-water reservoirs, were and still are being lost to urbanization, and groundwater became exposed to pollution by municipal wastewater that is collected in cesspools overlying groundwater aquifers.

The Department of Statistics classifies land use in Jordan as follows: grazing areas, 93.3% of total area; urban areas, 1.89%; forests and reforestation areas, 1.5%; water surface, 0.62%; and agricultural land, 2.69% (DOS 2003).

Legislative Background

International agreements and conventions to which Jordan is signatory are binding to the country. Jordan signed several global conventions aimed at preserving and protecting the global environment and contributing to the conservation of local ecosystems, including the Convention for Biological Diversity, Convention to Combat Desertification, Ramsar Convention to Conserve Wetlands, and World Heritage Convention, as well as the regional Convention for the Conservation of the Red Sea and the Gulf of Aden. The kingdom is also a subscriber to the IUCN and UNESCO Man and Biosphere Program. Ratification of these international conventions led the government to develop a number of relevant strategies in addition to Jordan's Water Sector and Agricultural Sector strategies. Following is an overview of the main environmental protection and sustainable development strategy documents and action plans that have been developed.

National Environmental Strategy and National Environmental Action Program

Jordan prepared its National Environmental Strategy (NES) in 1992 and National Environmental Action Program (NEAP) in 1995. NES emphasized the need to give priority to and urgently address issues of water salination, pollution, and depletion and maintain productivity of agricultural land threatened by encroachment of urban areas and soil erosion. NEAP was meant to guide the government's environmental policy and its impacts on investment decisions, articulate programs to implement such decisions, and formulate action plans to deal with Jordan's priority environmental issues.

Agenda 21

Jordan reinforced its commitment to the principles of sustainable development through its participation in the Earth Summit held in Rio de Janeiro in

1992, and in 2001, the kingdom's National Agenda 21 document was published. Agenda 21 clearly recognizes the need to preserve and protect Jordan's flora and fauna, unique habitats, and cultural heritage. It also identifies the most important resource management issue facing Jordanians as that of water and calls for integrated management of water resources, protection and management of potable water supplies, and maintenance of water quality and aquatic ecosystems.

National Biodiversity Strategy and Action Plan

In response to Jordan's ratification of the Convention for Biological Diversity in 1993, the National Biodiversity Strategy and Action Plan was launched in 2003. The NBSAP presents a number of specific actions geared toward the protection and sustainable use of biological resources, reduction of the impact of industry on biodiversity, promotion of integrated land use planning and water resource development, and achievement of a biodiversity-oriented society.

National Strategy and Action Plan to Combat Desertification

Jordan ratified the Convention to Combat Desertification in 1995, after which the Ministry of Environment embarked on a national effort to develop Jordan's National Strategy and Action Plan to Combat Desertification. The strategy has been finalized, and its launch is expected in 2006.

Water Strategy and Policies

In 1997, the Ministry of Water and Irrigation prepared Jordan's water strategy and policies relating to the management and use of municipal water and wastewater, groundwater, and irrigation-water resources. The strategy and policies call for tapping the full potential of surface-water and groundwater resources to the extent allowed by hydrogeological constraints, economic feasibility, and social and environmental impacts. They require the collection and treatment of wastewater for subsequent reuse in irrigation and groundwater recharge. The strategy builds on giving first priority of water use to basic human needs, and it emphasizes a priority to the sustainable use of previously developed resources, including those mobilized for irrigated agriculture in the Jordan Valley and other established uses.

Agricultural Sector Strategy

The Agricultural Sector Strategy, issued in 2001, stresses the role of the agricultural sector in protecting biodiversity, plant cover, and soil characteristics. It specifically highlights the role of organic farming in avoiding the dangers of chemical pesticides and their negative impact on biodiversity and associated environmental imbalance. The strategy also acknowledges the increased role of reclaimed water as a resource for irrigation, within the constraints of environmental health and technical safety.

Environment and Water Legislation

Before 1995, several ministries were responsible for monitoring and enforcing environmental quality and protecting natural resources. The Ministry of Water and Irrigation (MWI) was responsible for quality monitoring and protection of water resources; the Ministry of Agriculture for managing agricultural and grazing lands and forests; and the Ministry of Health for ensuring adequate public and environmental health conditions. This resulted in what was perceived as fragmented effort in terms of environmental protection, with a duplication of effort in some instances and insufficient effort in others.

In order to improve the situation, Jordan enacted its first law dedicated to environmental protection in 1995, establishing its first environmental protection agency. The General Corporation for Environmental Protection (GCEP) started as an autonomous institution under the umbrella of the Ministry of Municipal and Rural Affairs and the Environment and reported to the minister.

The improvement rate was sluggish, however, and the GCEP's lack of genuine autonomy from the ministry was seen as part of the reason. A new law, Environmental Protection Law number 1 for the year 2003, established a Ministry of Environment (MOE), which took over the responsibilities of the GCEP. The law entrusted the MOE with environmental protection in the kingdom, setting Jordan's environmental policy, planning and executing necessary actions for the realization of sustainable development, setting standards for and monitoring environmental quality, and most important, coordinating national efforts to preserve the environment. It also mandated that the ministry protect water resources and take actions to prohibit any activities that caused pollution to or degradation of water resources in the kingdom.

In line with the law, the Council of Ministers issued several regulations aimed at achieving the objectives, including nature protection, environmental protection against hazardous interventions, air protection, sea environment and shore protection, natural reserves and national parks, management of harmful and hazardous substances, management of solid wastes, environmental impact assessment, and soil protection. More regulations are on the way to be issued, including one on water resource protection. These regulations detail the environmental protection implementation measures and specify the actions that the ministry is authorized by law to exercise.

The role of the MOE is not to be confused with the role of the Ministry of Water and Irrigation and its affiliate authorities, the Water Authority of Jordan

(WAJ) and the JVA. Whereas the MOE's role is to protect water resources, the responsibility remains with the MWI for setting and implementing water policies in Jordan, developing water resources, and monitoring and enforcing water quality standards. Any responsibility assigned to any agency under a previous law in Jordan is amended by a subsequent law. In other words, the MOE's authority over environmental protection supersedes any previous assignment of such role to any other agency. This does not negate the responsibility of water agencies to take measures to prevent any threat to the water quality falling under their jurisdiction.

The WAJ, for example, is empowered under its law (law number 18 for the year 1988) and its amendments, among other empowerments, to specify requirements for the preservation of water resources and their protection from pollution and to regulate the uses of water, prevent its waste, and conserve it. The JVA is empowered under its law (law number 30 for the year 2001), among other mandates, to prohibit the pollution of water in the Jordan Valley, monitor water quality in the valley, and enforce quality standards.

In addition to the above two ministries, Jordan's public health law number 54 for the year 2002 gives the Ministry of Health the authority to control and monitor the quality of water pumped for domestic use. This ministry is also given the authority to check on sewage water networks, internal sanitary installations, and wastewater treatment plants to ensure conformity with health standards; it is empowered under that law to take any measure the minister sees necessary to protect public health.

Four main sets of water quality standards have been developed, based on local needs and international experience.

- *Reclaimed domestic wastewater (JS893:2002).* This standard sets maximum allowable limits for quality of reclaimed water that is discharged to streams, wadis, or water bodies or reused in irrigation. It provides for three allowable limits, depending on the end use: high quality, used for the irrigation of vegetables eaten cooked, parks and playgrounds, and sidewalks within city limits; a lower quality, for irrigating fruit trees, trees planted on the demarcation lines of the easement of highways, and green areas; and the lowest quality, for field crops, industrial crops, and forest trees.
- *Industrial wastewater (JS202:2004).* This standard sets quality requirements for the disposal of industrial effluents into wadis, rivers, or the sea; for groundwater recharge; or for reuse in irrigation. The standard states that industrial wastewater disposal or reuse shall not negatively impact the surface-water and groundwater quality in the area of reuse or disposal, and that irrigation reuse would have to take into consideration the guidelines of the United Nations Food and Agriculture Organization (FAO).
- *Uses of treated sludge in agriculture (JS1145:1996).* This standard is concerned with the agricultural application of sludge that results from wastewater treatment plants. It prohibits the use of untreated sludge for

agricultural purposes and specifies the conditions for the use of treated sludge on agricultural land.

• *Drinking water standards (JS286:2001)*. These standards set the quality requirements for water used for drinking and specify quality control for such water.

Each of the standards also specifies the mechanisms for monitoring and evaluation. The WAJ, through its laboratories and Quality Control Department, is the main institution responsible for monitoring various water-related standards, and it is implementing several diversified programs for quality monitoring of surface water, groundwater, and treated wastewater.

The Ministry of Health is in charge of quality assurance of municipal water supplies. It has its own laboratories, which are used mainly to ascertain the fitness of municipal water for drinking purposes. The Ministry of Environment, through a subcontract with the Royal Scientific Society, monitors the quality of drinking water, bottled water, domestic and industrial wastewater, surface water, and groundwater. The Jordan University Water Research Center performs quality analysis of water resources independently as part of its educational and research programs.

Water Resources and Quality Issues

In a country that faces acute water shortages, water quality is of vital importance, and protection of water resources assumes a high priority. Threats to water resources are mainly excessive abstraction, quality degradation, and depletion; the environmental impacts resulting therefrom are serious as well.

Sources of Water Quality Degradation in Jordan

A number of sources contribute to water quality degradation in Jordan. These include point and nonpoint sources of pollution, as well as quality degradation of groundwater aquifers, mainly in the form of increased salinity, caused by decreased recharge rates on the one hand and overabstraction from them beyond their sustainable yields on the other.

Point Pollution Sources. Wastewater treatment plants, meant to minimize environmental hazards, are major contributors to water pollution in Jordan, either through seepage into groundwater bodies or through direct discharge of treated effluents into wadis and surface-water streams. The As-Samra wastewater treatment plant, the largest in the country, receives about 75% of Jordan's wastewater and discharges its effluent into the Zarqa River. The base flow of the river consists almost totally of that effluent. The estimated flow of wastewater from treatment plants into side wadis in 2005 was 87 mcm/y, of which about

2 mcm/y infiltrated into the ground, contributing to the pollution of groundwater.

Cesspools, which constitute the localized treatment facilities in neighborhoods not connected to wastewater collection networks and in towns not provided such service, are another source of pollution. Wastewater collection and treatment service covers approximately 80% of households in Amman (as of 2003), and 65% in the whole of Jordan. This means that 35% of households in the country rely on local facilities for wastewater disposal. The percolating liquid wastes from these cesspools reach and pollute groundwater bodies, as was the case in 1999 with three springs serving the populations of Wadi Seer, Ajloun, and Jerash with municipal water. The estimated total contribution of cesspools to the groundwater aquifers in 2002 was set at 70 mcm/y; this will be reduced as connections to wastewater collection networks are increased.

Industrial wastewater is a third contributor to point-source pollution. Approximately 50% of Jordan's industries are located in the Amman–Zarqa conurbation, lying within the catchment area of the Zarqa River. Some of these industries are served by the wastewater collection network and cause difficulties in the treatment plant at As-Samra, while others not yet connected to the collection network dispose of their treated or untreated industrial effluents in the surrounding environment and dry wadis that run in winter. The primary pollutants from this point source are fat, oil, and grease (FOG) and heavy metals such as mercury, manganese, cadmium, chromium, and copper. No estimates of industrial contribution to the pollution of groundwater or surface water are available.

When exposed to rainfall, some or all of the chemical contents of solid waste landfills leach into groundwater resources, affecting their quality and fitness for use. Major solid waste landfills, such as the Russeifa, within the Amman–Zarqa conurbation, and Ekader, in northern Jordan, have been associated with groundwater quality degradation. In addition, solid waste is piled at a few scattered sites around industrial areas, waiting for pickup by trucks. These sites create similar hazards to groundwater. Better attention has been paid to curb this pollution at newly designed landfills. Fortunately, the Russeifa site reached its capacity and was closed, and a new landfill site, southeast of Amman and away from the crucial Amman-Zarqa aquifer, is now used.

Urban pollution consists of pollutants originating in the streets of urban areas, imparted by automobile fuel combustion, and solid waste scattered on the streets, parking lots, and other vacant land. Eventually, rain washes these wastes into surface-water streams. Moreover, the maximum discharge capacity of storm-water drains has been exceeded as a result of rapid expansion in urban areas, and they often overflow. To help drain the flooded streets, authorities open up the manholes of the wastewater collection system. The urban storm water thus collected, carrying pollutants, ends up in municipal wastewater treatment plants for which they are not designed. This practice exacerbates the

situation of already overloaded plants and degrades the quality of their effluents. No quantified estimate of the contribution of urban pollution has been
made.

Nonpoint Pollution Sources. Irrigation-water runoff carries pollutants such
as pesticides, herbicides, fertilizers, organic chemicals, heavy metals, and salts.
Irrigation water in excess of the soil saturation capacity finds its way down to
underlying aquifers, where the pollutants adversely impact the groundwater
quality. This phenomenon exists wherever farming is practiced in the recharge
domain of aquifers; it has been most observed in Dhuleil, Jafr, and the Jordan
Valley.

Rain-fed agriculture makes similar contributions to pollution. Animal
manure, fertilizers, pesticides, and other chemical pollutants are washed into
groundwater aquifers during the rainy season. After summer cropping, residues
of such chemicals and organic deposits are leached by rain during the following rainy season and wind up in aquifers.

Reduction of Aquifer Recharge. Urbanization in Jordan has caused negative
impacts on aquifer recharge rates. It has created changes in the natural water
balance of precipitation among surface runoff, soil water, and groundwater
recharge, to the disadvantage of the last. Overabstraction from virtually all the
groundwater aquifers, notably in the Jafr, Dhuleil, Zarqa, Azraq, and Mafraq
areas, has been causing increased salinity and a decline of the water table. As a
result, many wells in these areas have dried up, but the legally drilled ones are
replaced by deeper wells.

The Amman-Zarqa aquifer, for example, has lost a good part of its recharge
area in the west and south to urban expansion of Amman and satellite towns
around it. The result has been a decline in the recharge area and the diversion
to surface runoff of a good part of the precipitation water that historically
recharged the aquifer. No quantification of this effect has been made, nor has
it altered the practice of urban planners in order to preserve such areas to sustain their function. At the same time, pumping rates from groundwater in the
Amman-Zarqa aquifer have progressively increased unabated over the past
three decades. In attempting to quantify the impact of urban expansion on
aquifer recharge, analysts are faced with several factors that are difficult to
account for. One such factor is the identification of the effect of interannual
variability of rainfall; another is the contribution of irrigation return flow,
including flows from house gardens and green areas in the recharge area.

Table 5-1 summarizes water pollution in Jordan by source of pollution or
quality degradation, most affected sites, and most affected type of water.

Overabstraction of Aquifers. Jordan's rapid population and economic growth
has naturally created a corollary increased demand for water. As sudden as the

TABLE 5-1. Summary of Water Pollution in Jordan

Source of pollution	Site	Affected water type
Wastewater treatment plants	As-Samra, Mafraq, Ramtha, Baqaa, Irbid, Harash, Karak, Salt, Tafileh, and Aqaba	Surface water and groundwater
Cesspools	Amman, Irbid, Ajloun, Jarash, Azraq, Zarqa, and Russeifa	Groundwater
Industrial wastewater	Amman, Zarqa, Russeifa, and Baqaa	Surface water and groundwater
Irrigation water runoff	Jordan Valley, Dhuleil, Azraq, and Shobak	Groundwater
Agricultural pollution	Jordan Valley, Azraq, Dhuleil, and Shobak	Groundwater
Solid waste	Russeifa, Ekeider, Salt, and Madaba	Surface water and groundwater
Saltwater intrusions	Azraq and Dead Sea	Groundwater

Source: Salameh and Bannayan 1993.

influx of population has been, the additional water demand was fulfilled through increased abstraction from existing wells or new ones drilled in aquifers already in use. Agricultural abstractions from groundwater have especially increased in the Highlands, with a short time lag behind the expansion of irrigated agriculture in the Jordan Valley. In 2000, about 2,500 operational groundwater wells abstracted water for all purposes beyond the sustainable yields of aquifers. Table 5-2 shows the quantity of renewable and nonrenewable uses of groundwater from Jordan's basins.

Even though agriculture is the majority user of groundwater, it is industry that is most reliant on groundwater as a main water resource. Groundwater forms 95% of total water used for industry, mostly in the form of dedicated wells legally drilled by the industries.

In order to fully understand the implications of groundwater use, the Ministry of Water and Irrigation, supported by the Federal Institute for Geosciences and Natural Resources of Germany, developed a comprehensive groundwater model covering all major aquifers. Particularly pertinent data include the water budget, groundwater balance calculations for the entire country and individual subbasins, and figures for the calibration of rainfall–runoff calculations.

Using data from the groundwater model, further models and planning scenarios are being developed to aid in making close predictions concerning groundwater use. Socioeconomic and environmental considerations are being factored in, and scenarios for the use of groundwater with a focus on sustainability will be defined.

TABLE 5-2. Comparison of Groundwater Abstraction Rates, 2002 (mcm/y)

Basin name	Groundwater abstraction			Total renewable groundwater abstraction	Safe yield	Deficit renewable resources
	Domestic industrial tourism	Irrigation	Total			
Yarmouk	8.8	48.2	57	57	33	−24
Jordan Rift side wadis	22.4	4	26.4	26.4	31	4.6
Jordan Valley	8.7	25	33.7	33.7	20	−13.7
Amman-Zarqa	83	54	137	137	62	−75
Dead Sea	53	29.5	82.5	82.5	44	−38.5
Southern Desert	13.5	71	84.5	2[a]	3	1
Wadi Araba North	3.5	2	5.5	5.5	6.4	0.9
Wadi Araba South	1.3	4.3	5.6	2[a]	4.9	2.9
Jafr	15	10	25	25	7.7	−17.3
Azraq	23	35	58	58	34	−24
Sarhan	0.2	4	4.2	4[a]	5	1
Hamad	1	0	1	1	15	14
Total	233.4	287	520.4	434.1	266	−168.1

a. The remaining abstracted amount is from nonrenewable resources.
Source: MWI 2002.

Curbing Overabstraction

Jordan's water strategy and its associated policies emphasize the need to protect groundwater resources from quality degradation and pollution and give priority in use of the resources to municipal and industrial uses. The Ground Water Management Policy considers monitoring of groundwater resources and protection of recharge areas of aquifers as important factors in safeguarding their quality. To minimize overabstraction of aquifers, the policy stresses halting illegal drilling, metering all water wells, and using financial instruments for deterrence.

In line with these policies, several measures have been implemented to protect aquifers from degradation and overabstraction. These include the delineation of groundwater protection zones, preparation of groundwater vulnerability maps, establishment of a groundwater-monitoring directorate within the MWI, and issuance of the Groundwater Control Bylaw.

In 2002, the MWI, working with professional and trade organizations, environmental agencies, farmers, landowners, and other stakeholders, guided development and enactment of this bylaw, which was legislated to consolidate and strengthen state control of groundwater. It regulates groundwater well licensing, drilling, and water abstraction, and it set a tariff for water abstracted over and above the permitted annual abstraction rate.

This increasing block tariff structure takes into consideration the legal status of the well (licensed versus unlicensed), water quality (fresh versus brackish water), and geographical location, with special tariffs having been set for wells located in the Azraq Basin. Illegal wells are charged higher rates in order to encourage their owners to legalize them and obtain an abstraction permit. Such a permit would control the overall rates of abstraction from the aquifer, and the tariff is supposed to deter overabstraction. Tariffs on water from wells in Azraq abstracted in excess of the abstraction permit are higher than elsewhere to reduce overpumping from that particular aquifer.

Most well owners incur costs in addition to the set tariffs as the water table declines and water quality deteriorates. Analysis of cost components of groundwater use for the Amman-Zarqa Basin showed that they typically include increased energy costs of higher lifts, costs of well abandonment (such as from the temporary closure of farms), costs of well reconstruction and deepening, and costs associated with increased salinity. Under current water abstraction rates, these costs are expected to increase with decreasing water tables and increasing salinity.

The implementation of the provisions of the groundwater bylaw has been effective. By March 2005, 221 unlicensed wells were backfilled and 50 illegal drilling rigs seized, the rate of metering reached 93% of all licensed wells, and the MWI had received more than 510 requests for rectification of well legal status and to give a chance for more wells to rectify their status. The process of implementation continues to progress, although it has encountered some difficulties, especially in the Disi-Mudawwara area, where agricultural companies are using fossil water under contract with the government. Imposition of water tariffs, they claimed, violated the provisions of their contracts. Facing the MWI's insistence, the companies resorted to litigation in the courts. The new government that was formed in the spring of 2005 decided to honor the contracts signed with these companies and stopped charging them for water.

Initial results indicate modest success: the total groundwater abstraction for irrigation purposes from licensed wells decreased by 25 mcm in two years, from 202 mcm in 2002 to 177 mcm in 2004; and from unlicensed wells, from 40 mcm in 2002 to 36 mcm in 2004. Nevertheless, even though the groundwater bylaw was an important step toward managing the use of groundwater resources, the data are not sufficient yet to evaluate its long-term impact on abstraction, especially for irrigation purposes.

Water Resources and Ecosystems

This section examines the interrelation between water resource allocation and management and Jordan's main ecosystems. It looks at the main characteristics of ecosystems at Azraq, Amman-Zarqa Basin, Jordan River, Dead Sea and side wadis, and Aqaba and discusses their importance, associated land use, the way these ecosystems were impacted by decisions concerning water allocation, and the challenges facing them today. It also describes the few cases when water allocation decisions were made in favor of ecosystems.

Azraq Oasis

The Azraq Oasis is located northeast of Amman and was historically fed by Soda and Qaysiya springs in Azraq South and Aora and Mustadhema springs in Azraq North. The oasis is considered an important wetland and is protected by the Ramsar Convention, to which Jordan has been a signatory since 1977. The oasis has also been designated by Bird Life International as an important bird area, because it falls along a major migratory route.

The Azraq aquifer started to be used for municipal water supplies in the mid-1960s to supply the northern city of Irbid and environs. Drilling of illegal wells in the area began and saw an increase in the aftermath of the 1967 war, when law and order were hard to uphold and maintain. In 1978, a project to supply Amman with municipal water from Azraq sources was approved (Chapter 4), and pumping to Amman commenced in 1980, when 5 mcm/y were transported. The transportation rate increased to 22 mcm by 1990 and 25 mcm today, as compared with the rate of 12 mcm/y approved by the high-level meeting chaired by the king in 1978. The overdesign of the transmission system was done by the Amman Water and Sewerage Authority, in charge of water and wastewater only for Amman, not for the entire country. In addition to abstraction of water for drinking purposes, an estimated annual quantity of 25 mcm is abstracted for irrigation purposes. When an estimated quantity of 8 mcm abstracted for other uses is added, the total abstraction rate amounts to about 58 mcm/y, which is almost double the recharge rate estimated at 34 mcm/y (Table 5-2).

This overabstraction has not occurred without an impact: the water levels in the area surrounding the oasis have dropped by a few meters, resulting in the drying out of the discharges of Soda and Qaysiya springs feeding the oasis and increased salinity of the groundwater. In 1958, water salinity in observation wells in the Azraq Basin ranged between 340 and 970 mg/L. By the mid-1990s, salinity had increased to around 1,500 mg/L in deep wells and 10,000 mg/L in shallow wells. The groundwater in the central part of the basin is mineralized and sulfurous and generally of poor quality. Total dissolved solids concentrations currently range between 800 and 2,500 mg/L. In the western and northwestern rims of the basin, the quality is good, with total dissolved solids concentrations between 200 and 500 mg/L.

Drying out of the springs feeding the oasis led to drying out of vast surface areas of the oasis. A project undertaken by the Royal Society for the Conservation of Nature and supported by the Global Environment Facility/UNDP attempted to rescue and restore what could be saved, and a nature reserve has been established with an area of 12 km². The rescue effort resulted in partial improvement and the return of around 160 migratory bird species. In spite of all the efforts, however, the water flow into the Azraq Oasis is barely enough to support 10% of the wetlands that once existed (Gouede 2002; Azraq Oasis Conservation Project 1999). It is unfortunate that Jordan's environmental awareness and the resulting legislation and administrative arrangements came too late, at least for Azraq.

The recent groundwater bylaw places focus on agricultural wells located in the Azraq Basin, and the MWI, through its Water Authority, the WAJ, has decreased its abstraction from the Azraq aquifer for municipal purposes as a lead action. Availability of incremental supplies from the Jordan Valley makes this possible, and it is hoped that the WAJ will decrease its abstraction from the Azraq Basin as more supplies become available from additional sources. Along with the strict law enforcement over the Azraq wells, the reduction in abstraction will give the aquifer a chance to recuperate.

Amman-Zarqa Basin

The Amman-Zarqa Basin is the population magnet in the country and the center of its modest industry. It is home to more than 50% of the industrial establishments, including some of the largest ones, such as the Hussein Thermal Power Plant, Jordan Petroleum Refinery, Russeifa phosphate mines and consequently a landfill (finally closed in 2003), and the As-Samra wastewater treatment plant, which treats 75% of Jordan's wastewater. The basin also has a mix of tanneries, food factories, chemical manufacturers, and textile factories (Dorsch Consult and ECO Consult 2001).

The speed with which these industries sprang up often overtook the zoning for them in national plans, and the resultant shortages in public utilities led many industries to dispose of their wastes by simply dumping them into neighboring wadi beds. Subsequently, the government prohibited the disposal of untreated industrial wastes into the environment and exercised control on the polluting inputs of industries. Production of nondegradable detergents, for example, had to give way to degradable ones to better control the quality of the effluent of the As-Samra treatment plant; many industries were required to install units to pretreat their wastewater before disposal into the collection networks and delivery to the treatment plant.

The blending of the effluent with floodwater impounded by the King Talal Dam has further improved the effluent quality. The long travel of the effluent flow in the Zarqa River bed exposes it to further natural treatment through sunshine and aeration. Decreases in BOD_5, COD, TP, and TN are observed as water flows from the As-Samra outlet to the dam through the upper Zarqa River.[1] The raw effluent entering the plant since 1990 has not conformed to the design criteria meant to produce quality effluent, especially in respect to the daily rate of inflow.[2] Upgrading of the As-Samra wastewater treatment plant, currently under way, is expected to further improve the quality of treated wastewater by 2007.

A study (Al-Jundi 2000) targeting pollutants in the Zarqa River sediments showed elevated concentrations of heavy and trace elements, particularly zinc, chromium, arsenic, vanadium, cobalt, and zirconium, corresponding with the locations of industries along the river. These were attributed to discharges of industrial waste, sewage, and effluent from pretreatment facilities.

Because of population and industrial concentration, groundwater quality of this basin is the most vulnerable to quality degradation of any groundwater basin in Jordan. Salinity of many wells and springs has increased substantially to levels far above the standard value for domestic and even agricultural uses. Comparing water quality monitoring records for the periods 1985–1989 and 1995–1999 shows that among 53 wells in the Amman-Wadi Sir aquifer B2/A7, 40 have had a significant increase in electrical conductivity (EC), a measure of salinity. The average increase is 324.4 μs/cm, or 23.1% above the EC level of the earlier period.

The upper aquifer in the Amman-Zarqa Basin, the Amman–Wadi Sir, is also most affected by industrial activity, particularly in the Russeifa and Awajan districts. The main contaminant in this aquifer, which is the nearest to the surface, is nitrate, with concentrations exceeding 45 mg/L (Awad 1997). Elevated levels of selenium (above the World Health Organization's maximum permissible level of 10 ppb for drinking water) and other trace elements have been found in 6 out of 39 wells sampled, with a range of 0.8 to 111.5 ppb. The three wells exhibiting the highest levels of selenium were all located in proximity to the now closed Russeifa landfill (MacGregor et al. 1996). A recent study by Jiries and Rimawi (2005) found that concentrations of polycyclic aromatic hydrocarbons in groundwater collected from 4 wells located around the Russeifa solid waste landfill site ranged from 0.1 to 0.4 ppm, with an average value of 0.26 ppm, compared with an average of 9.2 ppm in Russeifa landfill leachate and 751 ppm in soil samples.

Industrial wastes in this basin have to be managed better. It may be advisable to build a central industrial wastewater treatment plant, on a build-operate-transfer basis, separate from the municipal wastewater treatment plant at As-Samra, whose effluent is being reused in irrigation.

Jordan River Basin

The Jordan River system is an important wetland area in the Middle East because it maintains many globally valuable species, such as the brown fish owl, common otter, Arabian leopard, rock hyrax, freshwater turtle, several endemic freshwater fish, freshwater snake, and many other endangered species. It is also an important migratory bird route, with an estimated 1 million birds passing annually through this narrow corridor, such as black and white storks,

Dalmatian and common pelicans, kingfishers, herons, sandpipers, shanks, francolins, and other globally threatened waterfowl.

The main problems of the Jordan River Basin are the diversion of its entire freshwater base flow upstream for irrigation and other uses and the diminished flood inflow below Lake Tiberias. In fact, the Unified Plan worked out by the U.S. presidential envoy in 1955 (Chapter 10) allows the diversion of the entire freshwater resources of the basin before they get to the Lower Jordan below Lake Tiberias.

The water quality in the Lower Jordan has to be rehabilitated. This can be done only by revisiting the water-sharing plan and the uses of each share. Environmental integrity of the river has to be restored for the preservation of its ecosystem. The importance of the river to human heritage has gained recognition after the discovery of the place of baptism in Bethany on the East Bank of the river. The rehabilitation of the Jordan waters below Lake Tiberias was an important feature of the Water Annex of the Jordan–Israeli Peace Treaty (Chapter 10).

According to JVA monitoring data at nine stations along the Jordan River, the water shows very high salinity in the summer months (average EC = 4,480–6,663 µs/cm). Boron concentration is occasionally over 10 mg/L.

The quality of Yarmouk water diverted to the King Abdullah Canal in Jordan is showing signs of deterioration. Shortly after its Yarmouk diversion tunnel sees daylight, the canal receives water carried by a second, 12-km-long canal from a groundwater well field at Mukheiba, followed shortly by water inflow from the Lake Tiberias reservoir in Israel. About 3 km downstream, the King Abdullah Canal receives water from Wadi Arab Reservoir, and after 65 km, immediately after the canal crosses the Zarqa River, the flow from the King Talal Dam joins it.

Input from the dam has a significant impact on water quality in the canal downstream, as illustrated by Figures 5-1 and 5-2, which show EC and nitrate (N-NO$_3$) concentrations for the year 2002 (ECO Consult and PA Consulting Group 2003). The figures indicate seasonal variations and compare the concentrations to the restriction levels for use in irrigation according to FAO guidelines (Ayers and Westcot 1985). The numbers along the horizontal axis correspond to sampling locations. Location 107, about 200 m from the inflow point of dam water into the canal, shows a clear jump in both EC and N-NO$_3$ values, especially during the summer months, when Yarmouk water does not cross the Zarqa River siphon and the entire flow of the canal downstream consists of King Talal Dam water.

Upstream from the Zarqa River inflow into the canal, the water quality is much better. This water is pumped for municipal uses in Amman via the Deir Alla–Amman Project (Chapter 4). However, the Yarmouk River bed and banks, in certain locations, consist of exposed shale and phosphate rocks, which are scoured by the river flow, so the water carries some phosphate in suspension.

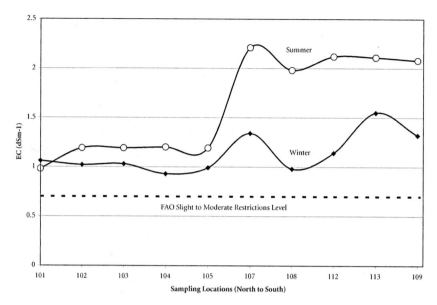

FIGURE 5-1. Variation in EC during Summer and Winter Seasons across the Canal
Source: ECO Consult and PA Consulting Group (2003).

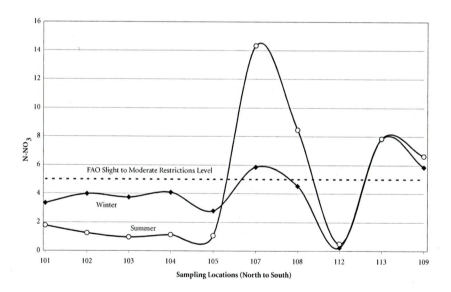

FIGURE 5-2. Variation in N-NO$_3$ during Summer and Winter Seasons across the Canal
Source: ECO Consult and PA Consulting Group (2003).

The fertilizing qualities of the phosphate cause algae to grow quickly during the hot summer months, causing problems for the irrigation system screens and the Amman municipal water treatment facilities at Zai. The disposal of untreated wastewater from some Syrian towns in the Yarmouk Basin augments the fertilizing power of the phosphate. The water transported from Israel has been known to carry other fertilizers, especially in the summer of 1998.

Groundwater quality in the Yarmouk Basin is variable, and concentrations of dissolved solids range from 300 to 500 mg/L where aquifers are under water table conditions, as is the case in the well fields of the Highlands in the basin. The same concentrations range from 550 to 660 mg/L where aquifers are confined, such as in the Mukheiba and Wadi Arab fields. In the Jordan Valley Floor Groundwater Basin, increases in chloride and nitrate concentrations appear to be related to water level fluctuations. In parts of the basin, a major concern is a rising water table because of irrigation return flows, which could potentially result in increased water salinity. In the southern part of the basin, water is slightly brackish, with chloride concentrations ranging from 700 to 1,850 mg/L; in the northern part of the basin, the water is somewhat fresher, with concentrations ranging from 500 to 1,000 mg/L.

Dead Sea and Side Wadis

The Dead Sea is an internal lake, naturally fed by the Jordan River and its tributaries and by direct inflow of the runoff of side wadis, the most important of which on the Jordanian side are Wadi Mujib and Wadi Hasa. The shore of the Dead Sea forms the lowest dry contour on earth and is important for the uniqueness of its natural environment and its historical and biblical significance.

The surface level of the Dead Sea has been dropping because of the cumulative effect of two factors: water inflow into the sea has been diminishing, while evaporation from its surface has intensified. Water budget studies indicate that the average water inflow, surface and subsurface, into the Dead Sea has decreased significantly over the past 30 years. Before the establishment of industries here, the total surface inflows averaged 1,670 mcm/y, 69% of which came from the Jordan River. In recent years, this average surface inflow has decreased to a mere 407 mcm/y, 58% of which is discharged through the Jordan River. Similarly, inflow of groundwater from the Jordanian side to the Dead Sea went down from an average of 220 mcm/y to 140 mcm/y within the same time period (Salameh and El-Naser 2000).

The result has been a drop in the Dead Sea level, from 392 m below sea level in 1920 to 407 m in 1990, and down to 416 m in 2003. Concurrently, the surface area of the Dead Sea shrank from 1,050 km^2 in 1920 to 634 km^2 in 2005, and the geometry of its coastline changed as well.

The government's policy on this issue has been to import water to the Dead

Sea from the Red Sea. Its negotiators in the bilateral peace talks with Israel succeeded in promoting this option, and a separate article was included in the peace treaty between Jordan and Israel on this matter (Chapters 1 and 10).

Water quality of the side wadis discharging directly into the Dead Sea is being monitored, as all of them are put to beneficial uses. Some are used in industry (Numeira, part of Hasa, and part of Mujib), others for municipal purposes (part of Mujib, Zara, and Zarqa Ma'een), and still others for irrigation (Ibn Hammad, Karak, Isal, Hasa, Feifa, and Khneizeera). These wadis can be categorized into two groups according to water salinity: a slightly brackish water group from Wadi Zarqa Ma'een down to Wadi Mujib, with EC values ranging between 1,500 and 3,000 μs/cm, and a freshwater group from Wadi Ibn Hammad down to Wadi Khneizeera, with EC values averaging 1,000 μs/cm or lower. No abnormality is noticed in other water quality parameters, because the catchment area of each of the side wadis is not heavily populated.

Water Allocation for Ecosystems

Although water allocation policies ignored the ecosystem requirements for a long time, the environment has scored successes in obtaining water in recent years. The diversion dam on Wadi Mujib, meant to divert its entire regulated flow from a point upstream of its discharge point to the Dead Sea, is a case in point. As presented in Chapter 4, such diversion would have jeopardized the sustainability of unique aquatic life in the wadi water downstream of its location and upstream of the discharge point in the Dead Sea. Additionally, the dam would have been built in the heart of a nature reserve managed by the Royal Society for the Conservation of Nature since 1986. The wadi is an important bird area and is home to at least 10 globally threatened flora and fauna species. It is also one of the cleanest and least disturbed river systems that has perennial water flow.

Protests and subsequent campaigns by the society persuaded the Jordan Valley Authority to modify the project and move the diversion dam a considerable distance downstream. The Mujib Diversion Dam was completed in 2004 at the agreed location to allow for water flow within the wadi (*Jordan Times* 1999; Khatib 1999; Shehadeh 2005).

Another case in point is the voluntary measure taken by the Ministry of Water and Irrigation to reduce pumping of water from the Azraq Basin to Amman after additional supplies were secured from other sources. The reduction is expected to increase as more water becomes available.

Yet another measure was the construction of the Wala Dam to recharge the base flow of Wadi Hidan after pumping from its groundwater reservoir to Amman for municipal uses had caused it to dry up. The water table has since bounced back, and the springs are now discharging fresh water into the wadi.

The giant Red Sea–Dead Sea Conduit is clear testimony to the priority given to ecosystems in Jordan and Israel. This linkage, as proposed by Jordan, has much less environmental impact than would linking the Dead Sea to the Mediterranean Sea, as originally proposed by Israel.[3]

Finally, the attention paid to migratory birds that have been attracted by the wetland created by the Aqaba wastewater treatment plant is further evidence of the place ecosystems have come to occupy in water resource allocation.

Aqaba

Aqaba Governorate has been transformed into the Aqaba Special Economic Zone (ASEZ), with plenty of investment incentives. A master plan that aims to promote industrial, business, and tourist activity in the area has been formulated for ASEZ. Even though environmental impacts and issues have been considered in developing the master plan, it is a challenge for the ASEZ Authority to protect and maintain the quality of Aqaba's fragile environmental resources in relation to the planned economic activity.

Aqaba's environmental resources include both land species and marine habitats. In the areas surrounding Aqaba, mammals that have been reported include the red fox (*Vulpes vulpes*) and Ruppell's fox (*Vulpes rueppelli*) in the desert and sand dunes, hyena (*Hyaena hyaena*) in rocky mountainous areas, and gray Arabian wolf (*Canis lupus*) in Rum, Batin Al Ghul, and Disi. The Arabian wildcat (*Felis silvestris tristrami*) occurs in vicinities near the road adjacent to Ras El-Naqab. A large number of snakes are also found here, mostly in the Wadi Araba region. These are mainly Afrotropical species, including the desert viper (*Cerastes cerastes*), Arabian saw-scaled viper (*Echis coloratus*), and desert cobra (*Walterinnesia aegyptia*).

Aqaba's marine coastal ecosystem is unique, with its sand and mud areas, rocky outcrops, coastal lagoons, fringing coral reefs, scattered coral heads, and several endemic species. Approximately 12% (around 80 species) of mollusks and a similar proportion of echinoderms occurring in the Gulf of Aqaba may be endemic. Only 15% of the gulf's amphipod species have been recorded as endemic to the Gulf of Aqaba and neighboring Red Sea. Several species of algae are also believed to be endemic. Of the 268 species of fish that have been recorded in the Gulf of Aqaba, 7 are recognized as endemic. Equally significant are 23 species of fish in the Gulf of Aqaba that are common elsewhere in the Indo-Pacific region but do not occur in the Red Sea. Together, these habitats constitute a delicately balanced, interdependent, and productive biological system that includes and sustains both coastal and pelagic fisheries (Mir and Abbadi 1995).

Aqaba falls on the major routes for seasonal bird migration, and its wastewater treatment plant's stabilization pond system formed a wetland where migrating birds take rest. The plant is currently undergoing rehabilitation and expansion and will employ extended aeration treatment techniques to produce better-quality effluent. The existing stabilization ponds, however, will be maintained to perform the environmental role they have been playing for migratory birds, which are an attraction for bird-watchers. In the process, the ponds will be used to regulate the flow of effluent to serve the existing irrigated areas near the plant. The policy of the Ministry of Water and Irrigation has been to prevent any effluent disposal into the gulf.

Wastewater Treatment and Reuse

Wastewater treatment plants in Jordan receive domestic wastewater that is relatively high in biological oxygen demand, a measure of the amount of oxygen consumed by microorganisms feeding on organic water pollutants under aerobic conditions, and total soluble solids compared with normally expected rates. This is because the per capita municipal water served, averaging 50 m^3/y, is below the desired average quantity served, and its salinity level is on the higher side of the allowable levels. Sixty percent of the water served to subscribers returns to the treatment plants; the rest is lost to leakage and to uses that do not send wastewater to sewers, such as gardening.

The As-Samra wastewater treatment plant is the largest of 19 plants operating in Jordan. In 2004, it received an average inflow rate of 179,000 m^3 per day, whereas its original design capacity was for 68,000 m^3 per day, a clear overload on the plant resulting in a degraded-quality effluent.

In 2002, the 19 wastewater treatment plants produced a total of about 81 mcm of reclaimed water, of which about 60 mcm were used in the Jordan Valley irrigation system, 5 mcm by private farmers around treatment plants under permits issued by the WAJ at a flat rate of 1.4 US cents per cubic meter,[4] 6 mcm without permits, and 2 mcm by industry. An additional 1.5 mcm infiltrated into the ground. About 92% of treated effluent was used that year. Treated wastewater amounted to 13.7% of the total water used in irrigation (517 mcm) and about 9% of the total water allocated for all uses (812 mcm).

Impacts on Agriculture

The collection and treatment of wastewater provide urban areas with immense environmental, economic, and public health benefits. Reuse of treated effluent has its advantages and disadvantages to agriculture. Its positive impact lies primarily in augmenting available water resources so that agricultural fresh water can be diverted for use in municipal supplies. Another positive aspect is that wastewater naturally contains fertilizing ingredients. Its disadvantage is that it adds salinity to the irrigated soils, resulting in lower yields. Salinity of the effluent increases as the per capita share of municipal water decreases.

Reuse of effluent in irrigation can also create potential health risks if pathogens are present in the wastewater. The hazards can be minimized by using pressure pipe irrigation systems, which minimize exposure of agricultural workers and fruits to potentially contaminated water. Irrigation systems that use treated effluent in the Jordan Valley are all piped networks.

Effluent reuse had several adverse impacts on farmers. Water charges for effluent were the same as for the fresh water that farmers had been allocated, but which was diverted to municipal uses. Farmers also incurred greater costs to safeguard against the health and environmental hazards the reused effluent posed. The effluent had higher alkalinity than the solution of certain fertilizers and pesticides allows, and thus using the effluent may have neutralized them. In addition, farmers had to purchase more potable water than they would have if they had received their freshwater quotas. The advantage of free fertilizing capacity in the wastewater did not mitigate these adverse impacts, and it has been found that the willingness of farmers to pay the tariff for irrigation water diminishes with increased use of reclaimed water, and that their attitudes toward its use are negative.

All these factors combine to affect the potential of agricultural land value. Partial analysis of potential impacts on the value of land is difficult and provides only a rough outline of future scenarios. Hijawi (2003) and Majdalawi (2003) separately examined the impact of lower water quality on yields for farms on both sides of the Jordan River, under very simplified assumptions. Results indicate, as expected, a reduction in yield and in farm income attributed to the degraded quality of irrigation water. A lower return from the agricultural land translates into lower value compared with land that earns higher incomes.

Reduced Crop Yields

The middle Jordan Valley is irrigated with water from the Zarqa River most of the time and with fresher water available from the Yarmouk from November to March. The Zarqa River water is treated wastewater mixed with storm runoff impounded in the King Talal Dam reservoir. Irrigation with Zarqa River water started in 1980, when flows from the Yarmouk receded because of increased Syrian use in the dry months.

The basic figures show that the total planted area of the middle Jordan Valley decreased from 8,237 ha during the first period (1987–1990) to 7,767 ha during the second period (1994–1998). The average total area of vegetables was 5,568 ha for 1987–1990. This area decreased to 5,359 ha on average for 1994–1998.

A comparison of vegetable crop yields between the two periods shows that they decreased between 24% (potatoes) and 9% (broad beans). Yields of summer vegetables were affected more severely, being reduced on about 1,330 ha, 80% of the area planted with summer vegetables.

The data from the reports of the Department of Statistics show a slight reduction in the total area planted with fruit trees in the middle Jordan Valley,

TABLE 5-3. Reduction of Land Shadow Prices in the Middle Jordan Valley Irrigated with Treated Wastewater (US$/ha)

Activity	Vegetables	Fruits	Field crops	Total
For the period 1987–1990				
Actual planted area (ha)	5,568	996	1,673	8,237
Shadow price of land	1,109	1,354	92	
For the period 1994–1998				
Actual planted area (ha)	5,359	786	1,622	7,767
Shadow price of land	721	1,052	43	
Reduction in land shadow price	−388	−302	−49	
% of reduction	−35	−22	−53	

Source: Estimates by Amer Salman and Emad Karablieh.

from 0.993 thousand ha during the first period to 0.967 thousand ha in the second. Oranges, clementines, grapes, and bananas were the main crops in the region with no significant change. Crop yields decreased at rates between 42% (grapes) and 21% (lemons).

Shadow Prices of Land

The complexity of the structure of land pricing and the interrelated effects of water quality and quantity make it difficult to isolate the impact of wastewater reuse on land prices. Table 5-3 presents a comparison of figures on land use in the Jordan Valley from 1987 to 1990 and 1994 to 1998, allowing at least a first estimation of the development of shadow prices of land under changing conditions in water supply. An unambiguous assignment of effects to water quantity and quality is not possible, however.

The shadow prices of land in the middle Jordan Valley according to crop groups showed consistent decline between the two periods: shadow price for land cropped with fruit trees dropped by 35%, with vegetables by 22.3%, and with field crops by 53%. Changes in the water supply in terms of quantity and quality affected shadow prices of land negatively.

Policy Recommendations

We make the following recommendations for taking into consideration environmental issues in the management and allocation of water resources.

- Even though several strategies, laws, and regulations address the issue of water resource management and conservation, the role of water in the maintenance of ecosystems and the allocation of water for the needs of nature are not specifically addressed or emphasized, except in irrigation to combat desertification and land degradation. The role of water in the maintenance of ecosystems should be emphasized and valuation techniques applied so that the true value of the services these ecosystems render can be taken into consideration in any cost–benefit analysis of water allocation projects.

- Advantage should be taken of the relative concentration of industries in the Amman-Zarqa Basin. The proximity of industries and the high proportion of food industries allow for benefits of economies of scale in building and operating an industrial wastewater treatment plant.
- Command and control systems used thus far for the protection of water resources in the kingdom should be reconsidered with a view to integrating other technical and financial instruments to improve efficiency of controls and implementation of legislation.
- Monitoring and enforcement practices can be improved by establishing baseline data on environmental quality, especially within sensitive areas. A database of operating industries linked to a geographical information system will be helpful.
- Reuse of treated wastewater in irrigation is increasing, as will the agricultural sector's reliance on it, and thus encouragement is needed. The cost of reusing treated wastewater should be made substantially less than, say, groundwater abstraction costs. A database of the users should be helpful in follow-up and controls.
- The management of both solid and liquid wastes directly impacts water quality. Medical waste of hospitals and clinics in particular should receive special attention.
- Environmental groups should participate in the planning and discussion of water projects and in environmental impact studies in order to decrease the negative impacts of reallocating water from nature to other consumptive uses.
- Regional cooperation is essential to address environmental issues of regional resources, such as the Dead Sea, Jordan River, and Gulf of Aqaba.
- Public awareness campaigns should be continued and reinforced to protect and conserve Jordan's valuable and limited environmental resources. These include awareness of water conservation, protection of water resources from pollution, and protection of biodiversity.
- Environmental impact and strategic environmental assessments should be used as tools to evaluate the possible consequences of development projects and other policies on environmental and water resources and to mitigate any negative impacts.

Summary

Jordan has four distinct biogeographic zones and seven ecosystems—desert, Mediterranean, subtropical, Dead Sea Basin, Jordan River Basin, Gulf of Aqaba, and freshwater—each with unique characteristics and environmental properties. Land use policy is vested in the Ministry of Municipalities, in charge of the planning for cities, towns, and villages, with the exception of the greater Amman municipal zone and the Aqaba Special Economic Zone. The Ministry of Agriculture is responsible for zoning agricultural lands, rangelands, and forests.

National strategies and action plans were formulated for the environment, biodiversity, and combating desertification, in addition to the water strategy and policies and the agricultural sector strategy. Faithful implementation of these policies and plans will enhance the efforts to preserve biodiversity and natural resources. The establishment of the Ministry of Environment in 2003 centralized the responsibility for environmental protection and planning.

Hazards posed to water resources include point- and nonpoint-source pollution, reduction of aquifer recharge, and overabstraction from groundwater aquifers. Administrative and legislative measures were taken to curb overabstractions, and results show that the drop in abstraction over two years was about 25 mcm from licensed wells and 4 mcm from unlicensed wells, a modest result thus far.

Overabstraction from aquifers has adversely affected important ecosystems, such as the Azraq Oasis. It affected the Amman-Zarqa aquifer with visible degradation of its water quality, as well as the quality of effluent in the Zarqa River, in which the industries disposed of their liquid wastes.

The ecosystem of the Jordan River Valley has been adversely impacted by the diversion of freshwater resources for various uses by the riparian parties. This ecosystem was home to a diversity of wild life and migratory birds. Increased freshwater diversion will cause the water quality of the Jordan to further deteriorate. Sharing and use of the basin water have to be revisited to allow its environmental recovery.

Water quality in the King Abdullah Canal is variable; it is lowest after the canal crosses the Zarqa River bed, at which point water-inflows from the King Talal Dam are responsible for the degradation. Suspended phosphate solids in the Yarmouk have a strong fertilizer effect that increases the growth rate of algae, especially in the hot summer months, when the algae can become a menace to irrigation water filters and the water treatment plant serving Amman. Groundwater quality in the Yarmouk Basin and the Jordan Valley floor is variable too, with different concentrations of nitrates and salinity.

Inflows into the Dead Sea have been diminishing as a result of water diversion and increased evaporation brought about by the surrounding industries in both Jordan and Israel. Surface-water inflows dwindled to 407 mcm/y, down from 1,670 mcm/y during the years preceding the development efforts. Groundwater inflow from the Jordanian side decreased from 220 to 140 mcm/y. Government policy is to transfer water from the open seas at Aqaba to the Dead Sea at Safi to stabilize the Dead Sea level and preserve the local and regional environment. Water allocation procedures have started to consider ecosystems, as evidenced by several projects undertaken in recent years, including the Mujib diversion, Azraq wetlands, and Aqaba wastewater treatment plant effluent.

Reuse of treated wastewater effluent in irrigation has been expanding. Effluent water quality conforms to the national standards, and reused water in 1982 amounted to about 13.7% of all irrigation water used that year. Treated wastewater reuse has its advantages and disadvantages. Recommendations for the improvement of environmental policy are made.

References

Al-Jundi, J. 2000. Determination of Trace Elements and Heavy Metals in the Zarka River Sediments by Instrumental Neutron Activation. *Nuclear Instruments and Methods in Physics Research Section B: Beam Interactions with Materials and Atoms* 170(1–2): 180–86.

Awad, M. 1997. Environmental Study of the Amman-Zarqa Basin, Jordan. *Environmental Geology* 33(1): 54–60.

Ayers R., and D. Westcot. 1985. Water Quality for Agriculture. Irrigation and Drainage Paper 29, Rev 1. Rome: FAO.

Azraq Oasis Conservation Project. 1999. *Environmental Impact Assessment and Implementation of Ramsar Convention. Comprehensive Report on the Monitoring Plan for the Most Important Wetland Sites in Jordan.* Prepared for the General Corporation for Environmental Protection and United Nations Development Programme. Amman, Jordan.

Dorsch Consult and ECO Consult. 2001. Feasibility Study for the Treatment of Industrial Wastewater in the Zarqa Governorate: Technical Proposal. Prepared under the Mediterranean Environmental Assistance Program. Amman, Jordan.

DOS (Department of Statistics). 2003. *Jordan in Figures: Annual Report.* Amman, Jordan: DOS.

ECO Consult and PA Consulting Group. 2003. *Water Resources, Supply and Demand & Water Balance: Water Quality Issues Supplementary Report.* Prepared for the Jordan Valley Authority. Amman, Jordan.

Gouede, N. 2002. Restoring an Oasis Strengthens Communities in Jordan. *Choices.* New York: United Nations Development Programme.

Hijawi, T. 2003. Economics and Management of the Use of Different Water Qualities in Irrigation in the West Bank. In *Farming and Rural Systems Economics,* vol. 50, edited by W. Doppler and S. Bauer. Weikersheim, Germany: Margraf Verlag.

Jiries, A., and O. Rimawi. 2005. Polycyclic Aromatic Hydrocarbons (PAH) in Top Soil, Leachate and Groundwater from Russeifa Solid Waste Landfill, Jordan. *International Journal of Environment and Pollution* 25(2): 179–88.

Jordan Times. 1999. Conservationists to Bring Awareness to Schools with Release of New Teachers Guide to Reserves. Jan 12.

Khatib, A. 1999. Jordan Signs JD66 Million Agreement for Dam Construction Projects in the Southern Ghor Region. *Jordan Times,* Jan 16.

MacGregor, R., B. Smith, M. Cave, A. Parker, and J. Rae. 1996. Elevated Levels of Selenium in Well Waters in the Amman-Zarqa Basin, Jordan. Paper presented at Society for Environmental Geochemistry and Health, European Meeting, Imperial College. April 1996, London, England.

Majdalawi, M. (2003). Socio-economic Impacts of Re-use of Water in Agriculture in Jordan. In *Farming and Rural Systems Economics,* vol. 51, edited by W. Doppler and S. Bauer. Weikersheim, Germany: Margraf Verlag.

Mir, S., and M. Abbadi. 1995. *Genetic Resources of Fish and Fisheries in Jordan.* Amman, Jordan: United Nations Environment Programme.

MOE (Ministry of Environment). 2003. *National Biodiversity Strategy and Action Plan.* Amman, Jordan: MOE.

MWI (Ministry of Water and Irrigation). 2002. Ministry of Water and Irrigation files. Amman, Jordan: MWI.

Salameh, E., and H. Bannayan. 1993. *Water Resources of Jordan Present Status and Future Potentials.* Amman, Jordan: Friedrich Ebert Stiftung.

Salameh, E., and H. El-Naser. 2000. Changes in the Dead Sea Level and Their Impacts on the Surrounding Groundwater Bodies. *Acta Hydrochim Hydrobio* 28: 24–33.

Shehadeh, Y. 2005. Personal communication between Yehya Shehadeh, acting director general, Royal Society for the Conservation of Nature, and the authors. May 15.

Notes

1. BOD = biological oxygen demand (BOD_5 refers to the BOD obtained as a result of a five-day BOD test), COD = chemical oxygen demand, TP = total phosphorus, TN = total nitrogen.

2. In 1998, the loading of the treatment plant was about 165,000 m^3 per day of raw wastewater, whereas its design capacity was for only 68,000 m^3 per day. The plant was enhanced since then, but not enough to meet the overload.

3. The Med–Dead Israeli proposal would have conveyed Mediterranean Sea water from Haifa down to Beit Shean, desalinated it there, and disposed of the brine in the Jordan River to flow down to the Dead Sea. This would have exacerbated the environmental problems the Lower Jordan already has been experiencing. The project would have had other adverse environmental impacts as well.

4. A fils is the smallest unit of the Jordanian currency; 1,000 fils make a Jordan dinar, which is equivalent to 141 US cents. Conversely, a US cent is equivalent to 7.08 fils.

6

The Economics
of Water in Jordan

*Amer Salman, Emad Karablieh,
Heinz-Peter Wolff, Franklin M. Fisher,
and Munther J. Haddadin*

Water is essential for life, but it is scarce in many parts of the world, particularly in arid and semiarid regions. Even in some water-blessed countries, the uneven spatial distribution of water resources leads to water stress in some areas. Water is scarce in Jordan, and dealing with such scarcity is thus a major policy issue.

The state holds the water of Jordan in trust for its residents. The policy in this regard has been reflected in laws that stipulate that water is a "state property." In this sense, water belongs to every member of the society, giving each the right to it, but at the same time, it is the property of no one. The state, through the Ministry of Water and Irrigation (MWI), regulates water use and attempts to ensure that use by some does not negatively impact the resource or cause appreciable harm to others. The question is how this can be accomplished optimally.

The basic concerns of economic analysis are the allocation of scarce resources and the relation of the value of those resources to their scarcity and allocation. The fact that water is essential for human life makes water and its allocation critical, but that does not exempt it from the applicability of the principles of economics. There are differences, even open clashes, worldwide between those who would treat water as a private economic commodity and those who insist that it is a good with a critical social dimension. The proponents of the first notion advocate the transfer of water to private hands, whereas the proponents of the second defend keeping water within the realm of the public sector.

The standard answer given by economists to the question of how best to allocate a natural resource is that this should be done through private markets.[1] That answer is often correct, but only if certain conditions hold.

- The market for the resource must be competitive, with many small buyers and sellers.
- All social costs and benefits involved in the use of the resource must be private ones, so that they are reflected in the profit-and-loss decisions of firms and the economic choices of consumers.
- The rate at which future costs and benefits are discounted by society must also coincide with the private discount rate.

At least the first two of these conditions, and quite possibly the third, do not hold for water, and definitely not for water in Jordan. The country has only one water owner—the state—and the water infrastructure involves large investments, making an arrangement with many small sellers difficult to construct. Further, even if one could devise a scheme for placing water rights in the hands of many small sellers, there would remain good reasons not to do so.

The principal reason is given in the second condition above. Not all social costs and benefits associated with water use are private ones. On the cost side, for example, the extraction of water from an aquifer will lower the level of the water table, increasing the costs of extraction at other wells. Further, private water extractors will not consider the possibility that their actions, together with those of other private extractors, will result in overpumping, thus damaging the aquifer for later use and possibly ruining it altogether. Similarly, private water users will not fully take into account the effect their individual use has on the environment.

On the benefit side, private markets will not provide water for positive environmental purposes. For example, in Chile, where a system of private water markets has been introduced, the habitat of flamingos has been endangered. In the water systems of the Jordan–Israel–Palestine region, private water markets would fail to offset the effects of the falling level of the Dead Sea or even to take such effects into account at all.

Looking at it from a different perspective, society has an interest in seeing that all its citizens—even those who might not be able to afford water at free-market prices—have an adequate amount of water.

Beyond all this, the fact that many countries, including Jordan, provide water to farmers at subsidized prices implies that the governments involved consider water in the hands of farmers to have a greater value than the farmers' willingness to pay. Such a view may be because of environmental effects—the desirability of green open spaces, for example; the fact that the loss of agricultural employment would cause social unrest; or political issues. Whatever the reason, neither the MWI nor the economic analyst can ignore such considerations.

It should be noted, however, that at least these last two matters might be handled in the context of a free competitive market for water. Poor people could receive cash subsidies, and farmers could receive subsidies for farm output rather than a lower price for water. In the latter case, the efficiency of the choice of agricultural inputs would not be harmed; in the former, the choice of consumption goods would be made by the consumers.

Finally, as to the third condition, the state may well have more concern for the welfare of future citizens than do private buyers and sellers.

Free markets in water are thus not a good solution to the water management problem. Governmental involvement is essential. But the question now becomes that of how such involvement should be optimally designed. We discuss this at the end of the chapter, after a description of governmental water management and the water situation in Jordan.

Private-Public Partnership

As part of the economic restructuring program that Jordan has been implementing since 1990, the government has pursued a new policy of managing water supply and sanitation. Supported by loans from the World Bank, the government has awarded management contracts for the provision of municipal water and wastewater services and management of systems maintenance. In particular, a management contract has been awarded to a private firm to manage the water and wastewater systems of the Greater Amman area, and more are planned for other regions in the country. Under this arrangement, the contractor is required to take under its own management responsibility the staff of the Water Authority of Jordan (WAJ) that had been in charge of operation and maintenance of the subject area, and it has the right, after a specified number of months, to return to the WAJ any employees that it sees as unfit for the job.

In an attempt to attract private sector participation in project financing, the government has also considered the idea of build-operate-transfer (BOT) in the development of water resources.[2] Through such approaches, the government hopes to achieve several goals: expanding the coverage and improving the quality of service, generating resources to finance future investments, increasing economic efficiency, reducing government fiscal burdens, and introducing technological advances.

Partnership between the private and public sectors regarding water and wastewater has thus emerged as a promising way to improve the performance of public water and wastewater utilities, expand the coverage and improve the quality of service, increase operating efficiency, provide alternative mechanisms of financing investment in infrastructure, and reduce the burden on government budgets. It is unrealistic, however, to expect the private sector to overcome the inertia of government's institutional and operational inefficiencies and to make up for all shortages in investment. Privatization of the water resources has never been on the table.

The success of the private-public partnership depends on the regulation role that government plays. Protection of the public against monopoly, over-pricing of services, and deterioration of service is mandatory to win the support and confidence of consumers. On the other hand, a private concern will not function adequately to meet the desired goals if government intervenes frequently in its operations under the guise of regulation, nor will it work for a loss or little profit. Thus government regulations have to address the concerns of the investor, the contractor, and the public. It is expected that private sector participation will lead to improved service quality and expanded coverage. To keep costs to consumers within limits of affordability, the government has to control water tariffs and contribute enough funds toward capital investments to make water charges affordable.

Many of the initial successes have resulted from relatively simple management improvements that did not require large investments or sophisticated technologies. Private firms have shown a remarkable capacity to improve the operation of existing infrastructure within a short time. For example, under the management contract for the Greater Amman region, performance efficiency of manpower has improved, collection rates have increased, and unaccounted-for water decreased from 54% to 48%. This is not to give the impression that management by the private sector has done magic, but the results testify to the superiority of private sector handling of water and wastewater operations, known to require prompt responses to maintenance requests, supply of needed inputs, and mobility in staff and equipment. Under government (WAJ) management, rules such as those regulating procurement, raises for efficient staff, and working overtime do not promote improvements in performance efficiency. Additionally, the management contract is being funded by a loan from the World Bank, whereas the previous management by the WAJ suffered from chronic budget deficits that handicapped performance.

The success of structural reforms in the water sector depends on sustained, determined political commitment to implement them; the support of supplementary reforms in regulatory regimes; a realistic and efficient tariff structure; and a clear policy on subsidy and its mechanisms to provide quality service to the poor. Effective regulation is a necessary, but not sufficient, condition. It is the cornerstone of sustainable private sector participation. But the creation of a regulatory framework does not by itself guarantee effective regulation; rather, implementation of such regulations is what makes the difference. Government has to allocate financial and human resources to guarantee honest implementation through supervision, monitoring, and active follow-up. It continues to lead in contacts and negotiations with donor agencies, borrow funds for investments, and guarantee repayment of foreign loans extended to its departments. The MWI, with its constituent entities, has been in charge of water administration and management since its establishment in 1988. In 1997, the MWI finalized an investment program with the year 2010 as its horizon. The investment program gets amended as years pass by and the situation changes.

We discuss later the methods through which infrastructure projects can be evaluated and the water system optimally managed.

The Population–Water Resources Relationship

With Jordan's population continuing to rise, the gap between water supply and demand at current prices will rise accordingly.[3] By the year 2020, if current trends continue, the per capita share of free freshwater potential (surface water 838 mcm and groundwater 266 mcm) available for all purposes will fall from the current 196 m^3 per person in 2004 (5.63 million persons) to only 127 m^3 per person (8.65 million persons), putting Jordan in the category of countries with an absolute water shortage. Soil water (866 mcm), used only for agricultural purposes, would, if conserved, add around 100 m^3/cap, making the total per capita availability about 227 m^3 in 2020.

Government's attempts to deal with the scarcity problem have focused not only on supply management, including rationing of water service, but also on demand management measures and the adoption of a public information policy. Nongovernmental organizations (NGOs) also were mobilized in the informational campaign in an attempt to increase public awareness and participation. Despite all measures, the coming scarcity problem will remain a major challenge facing water managers and the country at large. It will be more intense if the planned projects are not implemented in time. So far, government has secured only 42% of planned investment for water and wastewater projects. Delays in implementation of the projects are possible for reasons of finance or other unforeseen factors. A fallback solution obviously will be at the expense of irrigated agriculture. It is recognized that it is easier to move goods around than it is to move water (Allan 1991). Food imports will have to compensate for any loss of irrigation water. It takes 1,000 tons of water, for example, to produce 1 ton of wheat. It is easier to move 1 ton of grain from Nebraska to Jordan than it is to move 1,000 tons of water to meet the water shortage in the country.

Because of the inability of the WAJ to supply municipal water on demand to its entire service area, which is the whole Kingdom of Jordan, water rationing in many parts of the country has become the practice. Industrial, commercial, and tourism establishments are charged higher prices for their water consumption than are average consumers (the tariff specifies the equivalent of US$1.41/$m^3$ to hotels and industries). These factors helped develop a trucking business for water abstracted from private wells, thus prompting overabstraction to take advantage of market demand, especially in summer. However, the government policy, currently being strictly implemented, charges high prices for every cubic meter abstracted from groundwater in excess of the permissible annual abstraction rate.

A key policy question in designing urban water policy and institutional reforms is that of the appropriate structure of water charges to ensure long-term sustainability of service. Water pricing touches on equity and the willingness of the consumer to pay (including the consumer's ability to pay). As such, it becomes a politically sensitive issue, and government will not surrender this matter to private firms managing water and wastewater services. In order to determine optimal pricing policies, estimates of demand and supply functions for water have yet to be made. Studies are needed to focus on understanding the nature of household demand for water and should attempt to express it in household demand functions. Specifically, the household sources of water supply should be characterized along with quality of water service, cost of water, and levels of water charges in relation to household income, and policy-related implications should be based on cross-section household survey data. Unfortunately, such household data have not been collected in Jordan. The effort to develop demand functions requires careful planning based on long-term data availability and equity in water allocation and use, taking into consideration various water rights, priority for reasonable domestic use, socioeconomic development imperative, and needs of agriculture, industry, tourism, services, and the environment. No such studies have ever been made for Jordan.

We discuss later in this chapter how water prices should be efficiently set.

The Dilemma of Unaccounted-for Water

About 97% of the WAJ service area (the entire country) is covered by piped water networks, a high percentage indeed. However, in terms of measures of efficiency—percentage of nonrevenue water, hours of water availability, and number of personnel per 1,000 connections—Jordan has a poor record. On average, the WAJ provides water for a duration of only about 8 to 18 hours per week. Manpower deployment efficiency is impaired by overemployment, as evidenced by the number of personnel per 1,000 connections, and financial operation has always been below the break-even level.

The most dramatic evidence of WAJ management inefficiency is the high percentage of nonrevenue water (NRW) or water that is not accounted for because of illegal connections, leakages, human errors in meter readings and processing, and other reasons. Nearly 50% of water produced by the WAJ is not billed or not accounted for.

This contrasts unfavorably with the situation in other countries. For example, NRW is only 8% in Singapore, one of the lowest rates worldwide, and about 30% in Bangkok, which is about the average among developing countries. The WAJ's efforts to reduce the high percentage of NRW have focused on the replacement of networks to minimize leakage and, it is hoped, illegal connections. Investment in this regard has been undertaken by the WAJ, assisted by concessionary loans from friendly donors. This program has brought recent modest reductions in NRW percentages.

Until these results began to be obtained, the additional water brought to the towns and villages of Jordan by heavy capital investments did not increase revenue. However, restructuring of the municipal water tariff, made effective in 1997, did bring in more revenue. In effect, law-abiding consumers have had to carry the burden of the WAJ's inefficiency, because they have been paying for the unaccounted-for water, whose loss occurred through no fault of their own. It is felt that more equity should be reflected in the water tariff, with the WAJ carrying the burden of its inefficiency.

The Irrigation Water Tariff

Water is a constraint on agricultural expansion in arid and semiarid countries, and Jordan is no exception. Renewable freshwater availability in the kingdom, including soil water that supports rain-fed agriculture, is on the order of 337 m^3/cap/y (2004), compared with an average of 1,700 m^3/cap/y needed to meet the municipal, industrial, and agricultural requirements (Haddadin 2003) meant to make agricultural exports offset agricultural imports.[4] Nonetheless, the charges for municipal, industrial, and agricultural water have been subsidized for decades.

Under current water policy directives, agriculture occupies third place in priority of allocation of new water, after municipal and industrial requirements, and first place in allocation of treated wastewater. Although less than 25% of arable land is irrigated, agriculture consumed about 63% of all available blue water in 2002.

In 1989, the irrigation water tariff in government-operated projects in the Jordan Valley did not cover the operating and maintenance costs, let alone the water's scarcity value. At that time, the incentive to adopt advanced on-farm irrigation systems was prompted more by increased agricultural yields and the expansion of irrigated land area than by the water charges. In 1995, the price was adjusted upward by more than double, and a block tariff was introduced. The revenues from water sales, when collected in total, do not come close to covering the costs of operation and maintenance.[5]

After upward adjustment of water tariffs for both agriculture and domestic use, and because of water stress, the efficiency of water use has increased. For example, irrigation efficiency in the Jordan Valley reached 70% in 2000, compared with 57% in 1994. The restructuring of water pricing has helped increase this efficiency. In order to ease transition to market pricing of water, the increase of water price to agriculture has been scheduled over a number of years. The average price to agricultural users rose from about US$0.0052/m^3 in 1989 to US$0.031/m^3 in 1995 and about US$0.04/m^3 in 2000 (JVA 2005).

The Extent of Treasury Subsidies to Water and Wastewater Services

Jordan's Treasury remains heavily indebted despite debt forgiveness, rescheduling, and government buyback of debt. The ratio of total debt to gross domestic product (GDP) stood at 105% in 2000 (although it had declined drastically from its 1991 level of 180%). External debt amounted to 84% of GDP that year. The ratio of debt service payments to exports of goods and services (liquidity ratio) also fell to 15% in 1999. However, debt-rescheduling agreements have left Jordan with an inflexible debt service profile that persists over the medium term. The WAJ fails to receive revenue for about half the water it supplies (best practice is considered to be less than 15%), and cost recovery is low because of losses and low water prices. This necessitates large government subsidies to the WAJ, exceeding 1% of GDP.

Costs and Revenues of Irrigation Water in the Jordan Valley

The Jordan Valley Authority (JVA) is responsible for economic and social development in the valley. In the 1990s, the JVA recovered an average of 65% of its operating and maintenance expenditure, but only about 30% of its total annual costs, including depreciation (details of annual cost and expenditure are shown in Table 6-3 below). In terms of costs and revenues per cubic meter, operating and maintenance (O&M) costs averaged US$0.023 and capital costs US$0.027, totaling US$0.050/m^3, while the revenue averaged US$0.015/m^3, or about 30% of the actual total cost, leaving about 70% of the cost, or US$0.035/m^3, subsidized by the Treasury.[6]

Cost recovery, funding, and commercialization issues have recently come to the fore. Maintaining and improving service levels, and introducing institutional changes to carry them out, have intensified the JVA's need for flexible and sophisticated analytical and policy tools for utility management and financial planning. Further adjustment of the irrigation water tariff in light of diminishing agricultural returns would seriously jeopardize the sustainability of irrigated agriculture in the country. Jordan's entry into international agreements with the European Union and its ascension to membership in the World Trade Organization, coupled with the liberal import policies associated with the economic-restructuring programs, have burdened indigenous agricultural production. Additionally, the traditional markets of Jordanian agricultural exports in the Levant have been restricted by political rivalries and competition from other producers. Unless farm income is improved through better marketing

TABLE 6-1. Current Irrigation Water Tariff in the Jordan Valley

Water delivered (m³/month)	Tariff (US$/m³)
0–2,500	$0.01
2,501–3,500	$0.02
3,501–4,500	$0.03
More than 4,500	$0.05

Source: JVA 1995.

TABLE 6-2. Tariff to Recover Operating and Maintenance Costs

Usage level (m³/month)	Tariff (US$/m³)
0–1,000	$0.01
1,001–2,000	$0.02
2,001–3,000	$0.03
≥ 3,001	$0.05

Source: JVA 1997.

outlets, any increase in agricultural cost of production will increase the debt burdens of Jordan Valley farmers. For the purpose of serving the Jordanian farmer, the government is looking into the establishment of a specialized company with private sector participation to expand outlets and improve competitiveness. The current JVA irrigation tariff system has been in place since 1995 and makes no seasonal, geographic, or water quality distinctions. It is structured in four usage block charges, as shown in Table 6-1.[7]

A tariff structure to recover the costs of operation and maintenance is shown in Table 6-2.

The cost of water is relatively high in Jordan because of its limited availability. Water for municipal and industrial (M&I) uses is either abstracted from deep bore holes in the Highlands or pumped from the Jordan Valley in the form of surface water to Amman and groundwater to Irbid. The average estimated cost of M&I supplies delivered to the consumer is US$1.12/m³, compared with an average revenue of US$0.637/m³ in 1993–2002. This necessitated a subsidy of US$0.478/m³.

Groundwater used by farmers in the Highlands, financed exclusively by the beneficiary, is not subject to any charges, save when the beneficiary abstracts from a well more water than allotted (150,000 m³/y), in which case the beneficiary has to pay for each cubic meter of the overabstracted quantity (the charges are shown in Tables 6-4 and 6-5).

It is important to note that such costs do not include the scarcity rent of the water itself—the opportunity costs of not achieving the benefits that the water would bring in other uses. An efficient system of pricing would systematically reflect such rents, as is discussed in a later section.

TABLE 6-3. Annual Costs and Revenues of Irrigation Water in the Jordan Valley

	Irrigation water quantity (mcm)	Annual O&M cost (US$ million)	Capital cost (US$ million)	Total cost (US$ million)	Revenue (US$ million)	Deficit/ subsidy (US$ million)
1990–1992	264.3	5.8	5.5	11.3	1.0	−10.3
1993–1995	252.6	4.8	6.1	10.8	3.9	−7.0
1996–2000	234.3	6.1	8.2	14.2	5.5	−8.7
Average	250.4	5.5	6.6	12.1	3.5	−8.6

Source: JVA 2005.

Ignoring scarcity rents (which should not be done) and looking only at direct costs, the total cost of consumed irrigation water consists of the sum of O&M plus capital costs. A study made for the JVA by a U.S.-based consultant and funded by Agency for International Development (AID) regional funds indicates that the annual growth rate of investments in the irrigation sector throughout the last decade of the twentieth century was estimated at 4.83% (Forward 1998). Total costs, broken down into capital costs and O&M costs for different periods, are shown in Table 6-3. Volumes of water supplied are also shown, as are the direct subsidies that irrigation water received.

In contrast to the situation in the Jordan Valley, agriculture in the Highlands is sustained primarily by water from groundwater wells. Most of the licenses (two-thirds) of these wells define maximum abstraction quantities, and water meters are installed on almost all the wells, except that half of the meters have been deliberately or accidentally put out of order. There has been growing concern about overabstraction, and only recently did the Ministry of Water and Irrigation mobilize political support to legislate for groundwater control (see Chapter 2). By that legislation, the charges to be levied for overabstraction are shown in Tables 6-4 and 6-5.

The charges for overabstraction undoubtedly affect the feasibility of agricultural water use when they exceed a certain ceiling set by a blend of free abstraction and the charges for overabstraction. Note that the "free" quantity does not mean no cost to the well operator; it just means that the government does not charge for that quantity. The operator does pay for the entire operating and maintenance costs and for the depreciation of equipment. The well owner or operator will adjust the area farmed in light of the direct cost of production increased by the charges for overabstraction.[8]

Costs and Revenues Associated with Water and Wastewater Services

During the 1990s, the low level of water tariffs made it impossible for the MWI to come closer to the long-term objective of the Urban Water and Sanitation

TABLE 6-4. Charges for Overpumping Water for Licensed Wells

Water pumped (m³/y)	Charges (per m³)
0 to 150,000	Free
151 to 200,000	25 fils (US$0.035)
More than 200,000	60 fils (US$0.085)

Source: WAJ 2004.

TABLE 6-5. Charges for Overpumping Water for Nonlicensed Wells

Water pumped (m³/y)	Charges (per m³)
0 to 100,000	25 fils (US$0.035)
101,000 to 150,000	30 fils (US$0.042)
151,000 to 200,000	35 fils (US$0.050)
More than 200,000	70 fils (US$0.098)

Source: WAJ 2004.

sector to finance its operating expenses and capital investments from its own revenue stream. In fact, the WAJ operations have generated annual deficits in excess of US$50 million in 1990 and, in the process, have reduced the WAJ's net worth from US$175 million in 1990 to zero in 1995. The WAJ's inability to generate sufficient surplus to finance its investment program resulted in its incurring large debt obligations. By 1998, the WAJ's cumulative debt obligations amounted to about 10% of the year's GDP. This prompted the government to provide significant direct assistance (of nearly US$706 million by the year 1999–2000) to bail out the WAJ from its debt burden. However, debt relief is not the only resource that the government has been providing to the WAJ. Others have included capital investments and transfers to cover operating shortfalls that averaged US$70 million annually from 1997 to 1999. The income of the WAJ in 1990–1999 is presented in Table 6-6.

Table 6-7 shows indicators of the WAJ's performance efficiency in 1993–2002.

Legally, the WAJ is an autonomous corporate body with financial and administrative independence. Its budget is not part of the government's own budget, but the government is its backer and guarantor. A huge cost will accrue to society of covering the annual losses, recapitalizing the WAJ, and paying off its long-term debts. The WAJ's annual deficit alone exceeds 1.2% of GDP.

Consumers' Ability to Pay for Municipal Water

The poor, who consume 20 m³ or less of municipal water per quarter per connection, were accorded due consideration in designing the municipal water tariff. According to the World Bank, poverty in the country increased from 3%

TABLE 6-6. WAJ Historical Financial Performance (US$ thousand)

	1990	1992	1995	1997	1998	1999
A. Income before depreciation (operating revenues less operating expenses)	−5,065	−7,250	−10,373	−4,039	14,562	21,949
B. Net operating income (A less annual depreciation)		−26,786	−38,792	−51,972	−54,942	−36,525	−35,707
C. Net income (B less annual interest expenses)		−50,767	−59,848	−83,021	−79,320	−64,405	−67,850
Working ratio (operating expenses/ operating revenues)	1.17	1.19	1.19	1.06	0.84	0.76
Operating ratio (operating expenses including depreciation/operating revenues)	1.89	1.99	1.93	1.79	1.39	1.39

Source: WAJ 2004.

in 1987 to 12% in 1997. Many Jordanian analysts, using various poverty surveys, estimate poverty to about double the rate reported by the bank's missions.

In 1997, under pressure from lending institutions, the government adjusted the tariff upward, increasing the rate for the lowest consumption block from US$0.28/m³ to US$0.49/m³ on average. Table 6-8 shows the average charge per cubic meter from the various blocks of tariff.

Table 6-9 shows the percentage of total household expenses, exclusive of wastewater collection charges, that water expense represents.

As wastewater charges are approximately 50% of the water bill alone, the average expense for water and wastewater together was on the order of 1.34% of total household expenses in 2003. This expense, however, does not reflect the total real cost of water service. The total operational expenses in 1999, for example, amounted to US$34.82 per capita, while the collection rate from consumers amounted to US$20.69. The average subsidy thus amounted to US$14.13 per capita. Compared with per capita GDP in that year (US$1,412), the cost of water and wastewater service, excluding capital cost, amounted to about 2.5% of the GDP, of which the consumer paid 1.46% and the government contributed the rest.

It is estimated that the average consumer can afford to pay about 2% of income for water and wastewater services. This means that the current tariff could be adjusted slightly upward to recover a higher percentage of the operational cost of water and wastewater services. To recover the entire annual cost, per capita, incomes would have to double—and probably keep increasing in the future as water costs increase. For full cost recovery, Jordan is in a race with time to achieve higher rates of economic growth and assure a balanced pattern of income distribution. Until then, water and wastewater services will continue to

TABLE 6-7. Performance Indicators of WAJ, 1993–2002

Item	1993	1994	1995	1996	1997	1998	1999	2000	2001	2002
Number of employees	6,714	6,900	7,330	7,570	7,414	7,460	7,762	7,869	7,709	8,006
Total employee salaries (US$ million)	19.99	21.38	23.33	25.79	26.76	27.36	27.80	29.86	33.26	30.88
Average monthly salary per employee (US$)	249	258	266	284	301	305	298	316	360	322
Quantities of water billed (mcm)	92.41	97.82	104.07	108.03	108.76	105.62	109.24	112.91	117.59	124.93
Total water return (US$ million)	33.35	38.06	42.47	45.44	50.93	69.31	66.20	79.55	79.21	79.57
Sewage water return (US$ million)	14.76	15.96	17.36	16.17	18.78	23.78	24.18	24.76	31.58	28.74
Total return (US$ million)	48.11	54.02	59.83	61.62	69.70	93.09	90.38	104.31	110.79	108.32
Employment cost (US$/m^3)	$0.22	$0.22	$0.22	$0.24	$0.25	$0.26	$0.25	$0.26	$0.28	$0.25
Average water return (US$/m^3)	$0.36	$0.39	$0.41	$0.42	$0.47	$0.66	$0.61	$0.70	$0.67	$0.64
Average sewage water return (US$/m^3)	$0.16	$0.16	$0.17	$0.15	$0.17	$0.22	$0.22	$0.22	$0.27	$0.23
Average total water return (US$/m^3)	$0.52	$0.55	$0.57	$0.57	$0.64	$0.88	$0.83	$0.92	$0.94	$0.87
Total water connections (thousands)	532	548	569	595	621	650	672	696	773	726
Total sewage water connections (thousands)	265	278	290	307	325	360	376	391	415	444
Total beneficiaries (thousands)	797	826	859	901	947	1,010	1,048	1,087	1,188	1,170
Connections per employee	118.7	119.7	117.2	119.1	127.7	135.4	135.0	138.2	154.1	146.2
Quantities of water pumped (mcm)	165	175	234	234	237	239	239	235	239	245
Quantities of nonrevenue water (mcm)	72	77	130	126	128	134	129	123	121	120
Percent of nonrevenue water	43.90	44.20	55.46	53.83	54.14	55.85	54.21	52.04	50.80	49.07
Population (millions)	3.993	4.1394	4.291	4.444	4.60	4.7557	4.90	5.039	5.182	5.329
Per capita billed water (m^3/y)	23.1	23.6	24.3	24.3	23.6	22.2	22.3	22.4	22.7	23.4
Per capita pumped water (m^3/y)	41.3	42.4	54.4	52.7	51.6	50.3	48.7	46.7	46.1	46.0
Per capita billed water (L/day)	63.4	64.7	66.4	66.6	64.8	60.8	61.1	61.4	62.2	64.2
Per capita pumped water (L/day)	113.0	116.0	149.2	144.2	141.3	137.8	133.4	128.0	126.4	126.1

Source: WAJ 2004.

TABLE 6-8. Development of Average Municipal Water Charges, 1980–2003 (US$/m³)

Water block (m³)	Block midpoint (m³)	1980– 1985	1986– 1988	1988– 1990	1990– 1996	1997– 1999	1999– 2001	2001– 2003
0–20	10	0.30	0.28	0.28	0.28	0.41	0.41	0.49
21–40	30	0.32	0.27	0.24	0.24	0.24	0.24	0.31
41–70	55	0.44	0.40	0.38	0.38	0.41	0.68	0.72
71–100	90	0.58	0.55	0.54	0.58	0.62	1.10	1.12
101–150	125	0.72	0.71	0.69	0.78	0.82	1.37	1.38
151–250	200	0.81	0.83	0.83	0.95	1.03	1.75	1.77

Source: WAJ 2004.

be subsidized. Any elimination of the subsidy will be at the expense of other obligations by households, and this would prejudice the standard of living and the quality of life.

Table 6-9 indicates that, with subsidies to water and wastewater upheld, there is room for increased charges from the beneficiaries. The percentage of expenses assigned to water and wastewater can be increased to 2%, as compared with the percentages indicated in the table. The government may find itself forced to adjust the water tariff upward to account for the increase in fuel prices enforced in 2005 and 2006.

Farmers' Ability to Pay for Overabstraction

We now turn to the question of the cost of groundwater abstraction and its affordability. This cost varies with location, crop, and technology. In open-field vegetable farms in the Highland areas, the estimated water requirement for one dunum (du), corresponding to an application rate of 4 mm/day/du during eight months of cropping, is 1,000 m³/du/y.[9] The allotment given to each well under the new Groundwater Bylaw of 2002, 150,000 m³, is sufficient for the irrigation of 150 du without incurring additional charges. If a farmer wanted to increase the planted area by, say, 75 du, his decision requires the payment of US$3,884 to the WAJ for the overabstracted amount. The additional cost raises the total cost of water, estimated at an average of US$0.22/m³, by 8% to US$0.23/m³. The increase eats up an average of 40% to 30% of the net profit that such a farmer would expect from this classical farm in a year of depressed prices, or 10% to 7% of the net profit in a good year, depending on the cropping pattern. The adverse economic impacts of charges for overabstraction increase with the size of the farm. However, the present policy of charging for overabstraction may reduce, but is not likely to eliminate, overpumping from aquifers. At current charges, farmers can afford the modest increase in the cost of production.

TABLE 6-9. Development of Household Water and Wastewater Annual Expenses

Year	Water expense as % of total household expenses			Wastewater expense as % of total household expenses			Household water and wastewater expense as % of total household expenses		
	Jordan	Rural	Urban	Jordan	Rural	Urban	Jordan	Rural	Urban
1980	1.06	1.30	0.94	0.53	0.65	0.47	1.59	1.95	1.41
1987	0.76	0.56	0.82	0.38	0.28	0.41	1.14	0.84	1.23
1992	0.59	0.51	0.61	0.30	0.25	0.31	0.89	0.76	0.92
1997	0.73	0.84	0.71	0.37	0.42	0.36	1.10	1.26	1.07
2003	0.89	0.62	0.98	0.45	0.31	0.49	1.34	0.93	1.47

Source: WAJ 2004.

Water consumption per unit land area differs with the environmental conditions of the farm. In the Badia region, for example, for an average farm of 150 du with 60 greenhouses, a case investigated by the authors, the total consumption of water reaches 180,000 m^3 per growing season, or an average of 1,250 m^3/du/season. This makes the cost of water US$0.23/m^3 for the farmer. In the Mediterranean environment region in the neighborhood of Amman, on the other hand, the cost to the farmer is on the order of US$0.56/m^3, where a dunum requires an average of 725 m^3 per growing season. In terms of cost per dunum, the comparison becomes US$290 in the Badia to US$410 around Amman, energy and land costs included. The saving in the cost of water per dunum in the Badia more than offsets the reduction in agricultural yield per dunum compared with the more moderate environment of the Mediterranean region around Amman. Following the enforcement of the Groundwater Bylaw, farmers will have to pay a charge of US$1,059 for overabstraction beyond the allowable limit of 150,000 m^3. Divided by the area of 150,000 du, the extra cost is US$7/du, or an overall increment of water cost of US$0.006/ m^3. Such an increase in the water price represents 3% of the net profit that accrues from owned farms and 14% from rented farms. The most probable response to the charges for overabstraction will be to increase water use efficiency, adopt automated systems of irrigation to reduce labor cost, and increase the yield per unit flow. Farming will become more intensive, and overabstraction will continue in the Badia.

The most important adjustment will have to be made by vegetable farmers. Indeed, to keep their farms profitable, farmers—users of wells—are decreasing the quantity of water they pump and reducing the area they crop. In the Governorate of Mafraq, for example, vegetable farms cover about 44% of cropped land (or 29,000 du), of which about 70% are operated by tenant farmers (approximately 20,000 du). If these farmers continue to use the quantities of water per dunum to which they have been accustomed, they will have to reduce their cropped area by one-third to stay within the free-water zone. If they improve water use efficiency to apply only 750 m^3/du instead of the usual 1,000 m^3/du, they still will have to decrease the cropped area by 11%. The production of vegetables could be improved by advanced farming methods, so total production may not even decrease. Although farming in the Mediterranean climate may be expensive because of land rental prices, the charge for overabstraction could make a difference in profitability. The structure of overabstraction charges may not prove successful in achieving the desired objectives in all cases, however.

Water Demand Management

Pricing is one of the most important measures for demand management to reach an efficient, sustainable, and socially acceptable use of scarce water resources. Water service can be sustained only if its cost is met; failing that, the quality of service will deteriorate. Cost recovery must be done either through water or water-associated pricing or through governmental subsidies reflecting the view that water has societal benefits that exceed private benefits. In particular, a policy that intends to provide water for free implies the need for subsidies from other segments of society to the water-providing agency. A water-for-free policy along with a run-down distribution system can often result in the powerful and rich getting water cheaply through self-financed pipe systems while poor people buy water at excessive rates or drink unsafe water (Liu et al. 2003).

A free-water policy also has other problems. Longtime subsidization distorts people's perception of water as a scarce and therefore valuable resource. Low water prices thus are likely to engender excessive use and waste of water, worsening an already tenuous situation. The most effective instrument to break this vicious cycle is to allow for water prices that recover the cost of water supply and enable financial sustainability.

We discuss efficient water pricing more generally in a later section. Here we note two points. First, as explained earlier, free-market pricing of water will generally not be socially optimal. And second, although the capital costs of water infrastructure must be met, it is generally not efficient to do this in the water tariff itself (even assuming that there is not to be a subsidy). This latter point requires some further discussion.[10]

Suppose first that a particular large piece of infrastructure has been built and will not be used to capacity for some time. For simplicity, assume that its operating and maintenance costs are zero. Then it cannot be optimal to increase the tariff for water use in an attempt to recover the capital costs of the infrastructure. To do so would be to reduce water consumption, even though it costs nothing to use the already built infrastructure. This would unnecessarily lower the benefits that can be brought by the water system while leaving costs unchanged. Of course, if operating and maintenance costs are not zero, then charging for them in the water price makes sense. Capital costs, however, should be charged for *in the water price itself*, provided that the increased usage taxes the capacity of the system.

That is not to say that capital costs should not be recovered. Rather, they should be recovered in a way that does not alter incentives for water use at the margin. This can be done through a system of connection charges, taxation, or other ways not directly affecting water usage.

Social considerations play a major role in the pricing of household water (a major component of the municipal water sector) and are also relevant in the industrial and agricultural water sectors. There are, however, substantial differences between requirements for household water and that used elsewhere. Agricultural production can often use water of lower quality than that used by households, such as recycled, brackish, or untreated surface water. In addition, the elasticity of water demand—the response to changes in water prices—is higher in agriculture than in basic domestic water demand. Another significant difference is that water supply to households, industry, and services must be extremely reliable, whereas the reliance of the agricultural sector on a dependable supply of water may not be as important, especially when water is to be used for low-cash crops. As a result, agriculture, although the main water-consuming sector, tends to be the one that offers the largest potential for adapting water requirements to changes in the water availability.

Jordan's agriculture is subject to considerable uncertainty as to water supply, but it also shows great flexibility because of the possible choices among a large variety of crops that can be grown in the same area. This flexibility is due to annual field crops, such as wheat and maize, which can be grown by using different amounts and qualities of water during different growing seasons. Planning methods to deal with such issues have been developed and used, but the sensitivity of agriculture to water remains an important issue for many countries in formulating water policies (Amir and Fisher 1999).

Affecting Agricultural Water Use through Price Policies

The government can use policy instruments to affect water usage. The two principal instruments are price incentives and quantity constraints.

Salman et al. (2001) evaluated the responsiveness of agricultural water demand to prices for water of different qualities—surface, brackish, recycled, and groundwater—in the Jordan Valley and the Highlands. The applied linear programming model allowed also for a subsequent estimation of water price elasticity. Results for the valley show that when the price of surface water rises (for example, from US$0.20/m^3 to US$0.25/m^3), a reduction in the irrigated area occurs (from 23,513 ha to 22,052 ha). This is because some crops, such as alfalfa, leave the optimal solution, as they are no longer profitable compared with the other crops. The quantity of surface water demanded is reduced by 13 mcm/y.[11] On the other hand, the use of other water types, mainly recycled, increases by 6 mcm/y, partially compensating for the decline in surface-water usage.

Table 6-10 shows the price elasticities—the percentage change in water quantity demanded because of a percentage change in water price—for each water type and for all water types together. The elasticities in the "All water" rows are with respect to the price of the given water type just above.

The demand curve for surface water in the Jordan Valley that is implied by the optimizing model is shown in Figure 6-1.[12]

The own-price elasticity of surface-water demand was about –0.04 at the actual surface-water price of US$0.049/m^3. This is a very low elasticity showing that, starting at that price, an increase of 1% in the price of surface water will decrease the quantity of surface water demanded by only 0.04%. This,

TABLE 6-10. Price Elasticities for Irrigation Water of Different Qualities in the Jordan Valley

Type of irrigation water	Price elasticity at actual price	Price elasticity at midpoint
Surface	−0.0414	−0.9068
All water	−0.0269	−0.4229
Brackish	−0.2930	−1.0052
All water	−0.0101	−0.0344
Recycled	−0.4272	−1.2117
All water	−0.0632	−0.0712

Source: Salman et al. 2001.

however, is almost entirely due to the fact that the actual price is so low that a 1% increase adds only about US$0.0005/m³ to the price, hardly affecting demand. By contrast, at the midpoint of the range of surface prices considered (US$0.575/m³), the own-price elasticity of surface water demand is about −0.91. This means that, starting at that price, an increase of 1% in the price of surface water (corresponding to about US$0.058) will decrease the quantity of surface water demanded by about 0.91%, so that demand is slightly inelastic. Because the demand curve is approximately linear, what matters is its slope. An increase of US$0.01/m³ reduces surface-water demand by about 1.4 mcm/y.

Still using the optimizing model, one can regress the total water quantity demanded on surface-water price, holding the prices for brackish and recycled

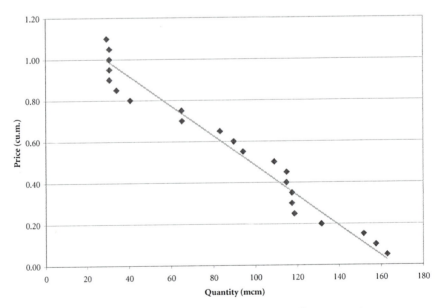

FIGURE 6-1. Surface Water Demand Curve in the Jordan Valley

TABLE 6-11. Water Quantities, Total Net Returns per Hectare, and Maximum Water Prices for Different Farm Types in the Jordan Valley

| | Farm types | | | | |
| | Vegetables | | | | |
	Plastic houses	Drip irrigation	Surface irrigation	Citrus	Fruits
Water quantity (m³/ha)	3,730	3,690	2,340	9,560	17,440
Average total net income (US$/ha)	2,458	777	565	4,520	8,785
Water expenses (US$/ha)	79	78	50	203	369
Maximum price of water (US$/m³)	0.68	0.23	0.26	0.49	0.52

Note: Figures, especially water prices, have been updated to 2003.

water constant. The overall water demand elasticity is –0.027 at the actual surface-water price of US$0.049/m³ but –0.42 at the midpoint price of US$0.575/m³ for surface water. Again, the demand curve (not shown) is linear, and total water demand decreases by about 1.3 mcm/y when surface-water price increases by US$0.01/m³.

The effect on overall water demand of increasing prices of brackish and recycled water is an elasticity of –0.01 with respect to the recycled-water price and –0.06 with respect to the brackish-water price at the actual prices. Even at the midpoints of the ranges studied, the elasticity is also small, being –0.07 with respect to the recycled-water price and –0.03 with respect to the brackish-water price.

In evaluating and discussing water demand management, it is worthwhile considering the water use and profitability of the different farm types. Salman (1994) presented and discussed the quantities of water allocated to each farm type and the corresponding profitability. The quantity of irrigation water used varies from one farm type to another according to the prevailing cropping pattern. The specialized plastic-house farms required 4,130 m³/ha of irrigation water, considerably lower quantities than did farms with fruit trees (17,440 m³/ha) and citrus (9,560 m³/ha). Farms that cultivated bananas and other fruits used an even higher rate of water per hectare. The affordability of the excessive use of this scarce production factor was mainly attributed to the low water price at that time (World Bank 1989), at up to US$0.009/m³; it has risen since then to US$0.0211/m³.

Table 6-11 shows the high variation of water consumption among the different farm types, as well as the maximum water price that each type would be able to pay. At this price, net revenues equal zero, provided that prices of other inputs remain constant. Although the specialized citrus and fruit farms had higher returns per unit area, they earned lower returns for the water used when compared with the specialized plastic-house farms. Maximum water prices were up to US$0.68/m³ for plastic-house farms, whereas prices did not exceed US$0.49/m³ for specialized citrus farms and US$0.52/m³ for fruit farms.

A general increase in water price would therefore overproportionally affect cultivation of vegetables, fruits, and citrus in the open field and support the tendency to change toward production in plastic houses. Expected consequences would include a loss in the variety of cultivated crops and an increased requirement for investments by farmers. That increased requirement might also engender further negative secondary effects, as small-scale family enterprises, which constitute the majority of farms in the Jordan Valley, might not be able to cope with the increased financial demands (Wolff and Nabulsi 2003). Benefit pricing, the coupling of water prices to the type of cultivation, has the potential for alleviating the consequences of the required adaptation of the current water prices to the real value of water as a scarce resource. Recent research indicates that, if this is seen as a desirable policy, water prices should be higher for fruit trees such as banana, apple, and citrus than for vegetables (Wolff and Nabulsi 2003).

Farmers' Response to Adjustments in Irrigation Water Tariffs

The Agricultural Sector Adjustment Loan (ASAL), extended to the Jordanian government by the World Bank, was first suggested in 1990, when the government was consulting with the International Monetary Fund on measures to help alleviate its economic and fiscal crises. The ASAL, in the amount of US$80 million, was approved in 1995, along with a US$7 million technical assistance loan. The objective was to promote efficient use of water resources through managing demand, deregulating markets, restructuring institutions, and improving planning and investment in the sector. Irrigation tariffs were raised by 150%, from US$0.008/m³ to US$0.021/m³ in 1997. Price controls for food, including fruits and vegetables, were abandoned; producer subsidies for wheat, barley, and tomatoes were lifted; and the land-lease market in the Jordan Valley was deregulated, allowing longer-term leases of land than the 10-year ceiling previously set.

Although agricultural exports grew by 11% per annum from 1995 to 1998 under the prevailing cropping pattern, it was expected that the ASAL would result in a shift from water-intensive crops to low-water-use crops. The results on the ground proved to be modest, however. Water meters installed on groundwater wells led to better knowledge of abstraction quantities, but they did not improve water conservation or income from water sales. Farmers were willing to pay more for water at a time when they were being squeezed by the decrease in subsidies.

The JVA raised the water tariff twice and in 1995 adopted an increasing block tariff structure designed to yield an average tariff of US$0.021/m³. Between 1995 and 2000, the average tariff was US$0.016/m³ based on billings and only US$0.012/m³ based on collections. But after 1997, the effective tariff declined by 10% as the JVA mitigated the adverse impacts of water shortages on farm incomes by forgiving or rescheduling collection as a form of financial relief. The tariff was levied on the total water flow per month in cubic meters per farm of 3.5 ha, in accordance with the schedule shown in Table 6-1.

It was anticipated that increased water tariffs would reduce agricultural water use, but this did not happen. Water allocation to farms has been made by the JVA on the basis of water availability and demand patterns, with quotas calculated for seasonal and perennial crops at times of shortage. Irrigation water shortages occur in lean years or when the Yarmouk River flow recedes in average and good years as a result of Syrian abstraction from springs and wells, as well as dams Syria had built in the catchment.

The first priority for water allocation under conditions of shortage is perennial crops, with the share per dunum scaled down to bridge the demand–supply gap. The justification for this allocation priority is the perceived need to protect the capital investment made in developing fruit tree orchards.

Next in priority come seasonal crops, with shares per dunum scaled down as well. The largest water shortage occurs in the dry months, but the supply increases in the winter months (November–April), when climatic conditions in Syria end the summer crops and their need for springwater, and precipitation in the lower catchment downstream of the Syrian dams contributes handsomely to the flow of the Yarmouk. Farmers of seasonal crops, facing the quota system for water allocation, opt to plant part of their crops during the early planting season in the Jordan Valley (August–September), expanding to the full area in November. Fruit trees normally are more profitable and owned by better-off farmers (owner–operators), whereas seasonal crops generally are planted by tenant farmers, and part of the profit goes to the owners. The quota system raises questions of equity, but these have been overruled by the economic considerations.

When the ASAL (World Bank 2003) validity was extended in 1993, a decrease in the planting of high-water-demand crops in favor of low-water-demand crops was projected, but the results on the ground were to the contrary. Bananas increased from 1,598 ha in 1994 to 2,060 ha in 2000. The irrigated area in the Highlands, dependent on groundwater that the ASAL meant to preserve by reducing overabstraction, increased instead, from 31,000 ha in 1996 to 42,000 ha in the year 2000, and most of the 11,000 ha expansion was in tree crops (8,900 ha), which consume more water per unit area of land.

It is not difficult to understand why this happened (Amir and Fisher 2000). The government was using two policy instruments at the same time: prices and quantity allocations.

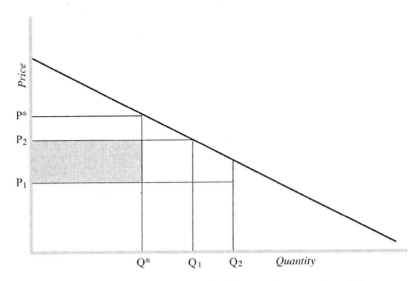

FIGURE 6-2. A Schematic Demand Curve for Above-Quota Water Quantities

Consider Figure 6-2. Here, Q^* is the amount of water allocated to the farm, the quota. P1 and P2 are the low and high prices, respectively, and Q1 and Q2 are the corresponding amounts of water that would be demanded if there were no quotas. Note that these are both greater than Q^*, indicating that the quota is a binding constraint on water use. In this circumstance, it is clear that raising the price from P1 to P2 has no effect on water usage. Instead, its only effect is to increase the payments made by the farmer for his allocated quota by $(P2 - P1)Q^*$, the area of the shaded rectangle in the figure. In other words, if the government's allocation policy was effective, then its price policy was not.

Optimal Water Management and Policy: The WAS Model

We now return to the issues considered at the outset of this chapter. We explained there that a free-market system for water management will not work. This is partly because water markets are unlikely to be competitive and largely because social benefits and costs from water do not coincide with private benefits and costs, but include other things as well. In this context, how should governmental intervention be guided?

The answer lies in a systematic study of the values placed on water and, most of all, the resulting value of water itself in different locations and at different times. In effect, this means building an economic model that reproduces the advantages of free markets while taking full account of those matters that free markets will not handle well.

Such a model has been built for Jordan by the Water Economics Project.[13] Here is a brief description of how this Water Allocation System (WAS) operates.[14] It uses the following inputs:

1. The country or region to be studied is divided into districts. Ideally, these should be as small as possible, but data typically exist by governmental entities, governorates in Jordan.
2. Within each district, data are collected on naturally occurring water supplies. These data include the location of sources, the annual renewable amount of each, and the cost per cubic meter of extraction and treatment.
3. For each district, demand curves are specified for each of three water-user types: households, industry, and agriculture. Such demand curves depend on factors that vary over time, such as population size, incomes, and agricultural prices; this corresponds to the fact that the WAS model is to be run for projected data for several future years.[15]
4. Infrastructure—conveyance lines, desalination plants, water treatment plants—is specified. This can be either existing infrastructure or possible future projects. In each case, operating costs and capacities are needed.[16]

After these data have been entered, WAS maximizes the net benefits that the country or region obtains from water subject to a large number of constraints:

1. The first set of constraints consists of the capacity limits on the infrastructure.
2. The second set consists of restricting water extraction from any source to that source's annual renewable flow; this can involve constraints across districts, if different districts can use the same underlying water source.
3. The third set consists of constraints on water use or prices placed by the user of the model, as explained below.
4. Finally, further constraints state for each type of water and each district that the amount of water used in the district cannot exceed the sum of the water extracted there plus the water imported into the district less the water exported from the district.

Although the constraints in the first three sets can be varied by the user, the constraints in the fourth cannot. Thus, for example, the user can permit overpumping of a particular water source or specify a particularly wet or dry year; he or she may also inquire as to the effect of changes in infrastructure capacity. But the user cannot change the physical facts embodied in the water constraints themselves.

We come then to the question of what is being maximized, of what is meant by net benefits from water. In the case of purely private benefits, this can be described as total gross benefits less costs. Total gross benefits consist of the amount that water users would be willing to pay for the quantity of water they receive. It is not hard to show that this can be measured by the area under the consumers' demand curves as water quantities go from zero to the amount actually supplied. Costs mean the total costs (excluding capital costs) of water supply and also an environmental charge where appropriate.[17]

The handling of the social benefits and costs that are not simply private benefits and costs can be thought of in two ways. The first of these is direct. Suppose, for example, that farmers are to receive a subsidy of US$0.10/m³. This is a statement that water in the hands of farmers is worth US$0.10/m³ more to society than it is to the farmers themselves. One can handle this by raising the agricultural demand curve for water by US$0.10, and then using the area under the altered curve to measure benefits. Unfortunately, for more complicated price policies, such as block rates, or for nonprice issues, such an approach is difficult to implement.

The second approach is easier. This amounts to permitting the user to impose constraints that directly reflect general policies. For example, the user can require that farmers receive water at a price US$0.10 lower than do other water users or according to a set block-rate schedule. Among other things, he or she also can set aside water for environmental purposes (or any other purpose), restrict the usage of treated wastewater, and require that certain groups of consumers obtain some minimum amount of water.[18] WAS then maximizes the objective function subject to these constraints that reflect public values and policy.

Now, as is generally true when something is being maximized subject to constraints, the optimal solution is accompanied by a system of shadow prices. Each shadow price is associated with a particular constraint, and each shows the amount by which systemwide net benefits from water would increase were the corresponding constraint relaxed by one unit. For example, the shadow price associated with a particular conveyance line capacity describes the systemwide change in net benefits that would be achieved were that capacity one unit larger.

But the most interesting shadow prices in the WAS solution are those corresponding to the water constraints themselves, the fourth set listed above. The shadow price of the water constraint in a given district shows the systemwide addition to benefits that would be achieved given (costlessly) one more cubic meter of water per year in that district. *These shadow prices are the true values of water in the different districts,* given the policy-imposed constraints involved and the rest of the input data used. Further, the shadow price of water at a source in situ is the scarcity rent of water from that source—the amount the country or region should be willing to pay to obtain an additional cubic meter of water from that source. The shadow price at any location that uses water from that source consists of the scarcity rent plus the per-cubic-meter costs of extraction, treatment, and conveyance to the user.[19]

If only private benefits and costs were involved, the water shadow prices would be the same as those arrived at by a free, competitive market. Note that they are not simply the marginal costs of extraction, treatment, and conveyance, but include a charge for scarcity value. Without additional government policies, such shadow prices would be the prices to charge to induce the efficient flows that would maximize the net benefits from water.

When the government has set price policies regarding water, the shadow prices will not be the ones seen by water consumers. In that case, they can be thought of as costs that the government is implicitly paying before reselling water to consumers at stated prices.

But shadow prices remain very useful even when the government decides that other prices should be charged. For example, the shadow price of water in a given district indicates the amount that it would be worth spending to obtain additional water there. This can be used to evaluate imports, new natural water sources, or desalination facilities. This brings us to the use of WAS to evaluate proposed infrastructure projects.

Evaluating Infrastructure to Avert a Water Crisis in Amman

Here we will concentrate on predictions for the year 2020 and the water problems in Amman. In this connection, we analyze a series of infrastructure projects that all have been undertaken, planned, or contemplated by the government. The discussion that follows is drawn from Chapter 7 of Fisher et al. 2005.

Figure 6-3 shows shadow prices for 2020 for the different governorates, assuming, contrary to fact, no changes in infrastructure after 1995 and no subsidization of water for agriculture. (Without the latter assumption, the water crisis in Amman would be at least as severe as depicted.)

These shadow prices show water crises in at least Amman, Zarqa, and Ajloun, with the shadow prices in Amman and Zarqa exceeding US$30/m^3. Such shadow prices for water are clearly unacceptable as actual prices and, with neighboring districts at much lower shadow prices, indicate a strong need for infrastructure improvements.

In Amman, the main infrastructure problem plainly involves getting water to the capital. In 1995, a conveyance pipeline carried 45 mcm/y of water from Balqa to Amman at an operating cost of US$0.22/m^3. Were no further infrastructure to be built, the shadow price in Amman would exceed US$30/m^3 by 2020, yet the shadow price in Balqa, in the Jordan Valley, would be only about US$0.16/m^3, and shadow prices elsewhere in the valley would be lower still. Plainly, the capacity of that pipeline will not be sufficient by 2020. Hence, either the pipeline must be expanded or other ways found to supply the capital.

Note that this is a problem not of water ownership, but of infrastructure. The shadow price of water ownership remains relatively low in the Jordan Valley despite the enormous shadow price in Amman. Because it is always the case in

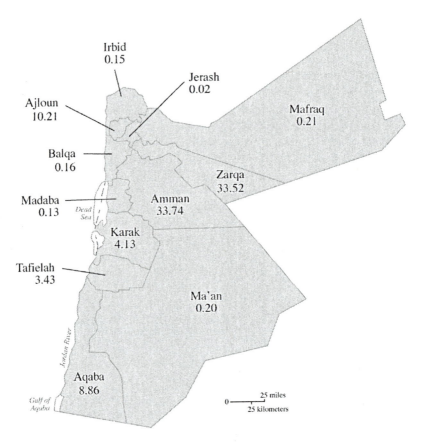

FIGURE 6-3. Baseline Shadow Prices in 2020 (US$)

the optimum solution of the model that, for conveyance from A to B, $p_B = p_A + t_{AB} = \lambda_{con}$ (where p_B denotes the shadow price at B; p_A, the shadow price at A; t_{AB}, the operating cost of conveyance per cubic meter between the two points; and λ_{Con}, the shadow price of conveyance capacity), the shadow price of the capacity constraint on the Balqa–Amman pipeline must be US$33.36/m^3 of annual capacity shown in Figure 6-3. This is the rate at which systemwide benefits would increase per cubic meter of additional conveyance capacity.

Here is an illuminating illustration. In 1994, when the Water Economics Project was in its infancy, Dr. Munther Haddadin, the editor of this volume and a later major participant in the project, was exposed to the proposed methods for the first time. He asked a rhetorical question: "If the two of us were lost in the desert east of Amman, what then would be the value of a bottle of water?" The answer is that the value of water in the desert would be very high indeed, but the value of water in the Jordan River would not change as a result. In such

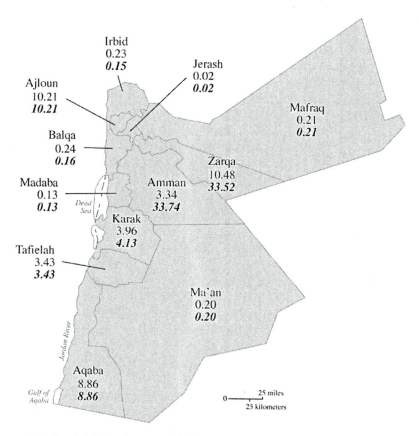

FIGURE 6-4. Shadow Prices in 2020 with (Upper Figures) and without (Lower Figures) Pipeline from Balqa to Amman Expanded to 90 mcm/y and Zarqa Ma'een Project Completed (US$)

a case, what is involved is a shortage of infrastructure to convey the water from the river to the desert, not a shortage of ownership of the resources.

Not surprisingly, therefore, the Jordanian government is expanding the Deir Alla–Amman pipeline from 45 to 90 mcm/y. In addition, the Zarqa Ma'een project would bring 35 mcm/y of desalinated brackish water from the Balqa district at a cost of US$0.47/m³; it starts operation in 2006.[20] Beginning with these changes in infrastructure, we see the immediate impact on Amman in terms of shadow prices, which drop from US$3.58 to US$0.47—the cost of the Zarqa Ma'een project water—in 2010, and from US$33.74 to US$3.34 in 2020, as shown in Figure 6-4. Note that shadow prices increase along the Jordan Valley from which the water is withdrawn. Irbid sees an increase from US$0.15 to US$0.23, and Balqa from US$0.16 to US$0.24. The scarcity of water in these governorates increases with the increased competition for water. With an overall gain in social welfare of approximately US$67 million per year in 2010, and reaching more than US$1 billion per year in 2020, the pipeline and desalination projects are clearly essential infrastructure.

In the discussion that follows, we assume the capacity of the Deir Alla–Amman pipeline to have been expanded to 90 mcm/y, and the Zarqa Ma'een project to have been constructed.

In addition to the increased pipe capacity from the Jordan Valley (Balqa) to Amman, another approach to alleviate the crisis would be to reduce intradistrict leakage. The government of Jordan already has plans to bring leakage down to levels of 15% by 2010. Although this reduction in leakage clearly lowers shadow prices in the crisis governorates, particularly in Amman, where the shadow price falls to US$1.92/m³, the values are still quite high in half of the governorates, especially Zarqa (US$6.92/m³) and Ajloun (US$6.08/m³). The total gain in social welfare from this reduction is an annual net benefit of almost US$220 million in 2020, suggesting that this could be a critical investment for Jordan over time.

In all of this, the shadow prices in several governorates adjacent to those in crisis—for example, Irbid, Balqa, Jerash, and Madaba—are much lower than those in Amman and Ajloun. This suggests the possibility of interdistrict conveyance. There are social limitations to these transfers, however, in that agriculture in the governorates with lower shadow prices is of great social importance, for employment as well as the aesthetic and cultural values associated with agriculture.

The Jordanian government is instead planning to use water from the Disi fossil aquifer to address the problem of persistent water shortages. It appears that pumping from this aquifer at a rate of 125 mcm/y is possible for a period of 50 years. A total of 70 mcm is currently being used from this system. An additional 55 mcm/y is added to the system in a new scenario, as well as a pipeline to Amman (initially of unlimited capacity, to let WAS determine the optimal size). Conveyance costs are estimated at US$1/m³.[21]

The results obtained from WAS show that the Disi pipeline would not be used at all in 2010 but would very clearly alleviate the crisis in Amman in 2020 more than does reducing leakage alone. However, the shadow price in Amman remains quite high (US$1.13/m³), and those in other districts are even higher. The net annual benefits for the Disi pipeline alone are approximately US$40 million in 2020. With a discount rate of 5% and a 20-year project life, the net present value is about US$500 million, which should be compared with the capital costs of the pipeline (estimated to be about US$600 million). This assumes no increase in population after 2020, so the actual net benefits are presumably higher.[22] Reducing leakage to 15% will have a more immediate impact on Jordan's social welfare, largely because it permits a net gain of 10% more water (as baseline leakage is 25%) throughout the country, whereas the Disi pipeline addresses one district's needs only. The volume of water that would efficiently flow through the new pipeline, according to our results, is almost 40 mcm/y in 2020.

But that is not the whole list of possible large infrastructure projects. A long-standing proposal has been to construct a canal with an annual capacity of 850 mcm from the Red Sea to the Dead Sea, known as the Red–Dead Canal or Peace Canal. The difference in elevation would be used directly to desalinate seawater through reverse osmosis and to generate electricity, and after pumping, it would provide much-needed fresh water to northern Jordan. In addition, the project would make it possible to stabilize the level of the Dead Sea. This would be environmentally beneficial and could enhance tourism and therefore boost the region's economy.

It is worth examining whether this development makes sense from the perspective of water needs in Jordan, with a particular focus on Amman, the district with the highest concentration of population. In 2020, with the increased supply of water from the Jordan Valley to Amman, a new pipeline from Disi to Amman, and leakage reduced to 15%, the shadow price in Amman remains at a relatively high US$1.08/m^3.

Assuming the Red–Dead project would deliver fresh water to Madaba, the water could then be transferred to Amman at an operating cost of US$0.22/m^3. As long as the marginal costs of the desalination did not exceed the difference of US$1.08 and US$0.22, or US$0.86, the project would likely be beneficial from the standpoint of social welfare. It also would likely lower the costs of desalination, because the required energy—a major component—would come from the canal's own hydropower plant. But using current costs (estimated at US$0.60/m^3, inclusive of capital costs), shadow prices in Amman could drop below US$0.80/m^3.

We assume here that the Red–Dead Canal is to be undertaken for reasons other than solely the production of desalinated water. The capital costs of the canal itself therefore should not be attributed (at least not in significant part) to the desalination part of the project. Furthermore, the Red–Dead desalination project would make it inefficient to transfer water from the Disi Aquifer to Amman, because the difference between the two shadow prices would not justify the transfer costs (US$1/m^3). This does not mean, however, that building the line from the Disi Aquifer would not be valuable if the Red–Dead Canal is to be constructed; quite the contrary. The transfer from Disi may well be needed between 2010 and 2020, while the more complex and time-consuming Red–Dead project is being approved and constructed.

Evidently, by embodying the economics of water, the WAS tool can be a powerful aid to water decisionmaking in Jordan. It can be used to evaluate not only the systemwide benefits of infrastructure projects, but also the systemwide costs (not just the direct governmental costs) of different water policies. Finally, as shown in Fisher et al. 2005, it can aid in water negotiations and help guide international cooperation in water.

Summary

Dealing with scarce water resources is a major policy issue. Water in Jordan is state property, with the government in effect holding it in trust on behalf of the people. Water is considered by some as an economic commodity and by others as a good with critical social dimension. This understanding affects the allocation of resources, with the first notion advocating water allocation through private markets. Three important conditions are necessary to support the first argument: water markets must be competitive, cost and benefits must be private, and future discount rates must be those adopted by society. These conditions are not met in Jordan, meaning that free market is not a good enough solution to the water management problem, and government involvement is essential.

Partnership between the private and public sectors regarding water and wastewater has thus emerged as a promising way to improve the performance of public water and wastewater utilities, expand service coverage and improve the quality of service, increase operating efficiency, provide alternative mechanisms of financing investment in infrastructure, and reduce the burden on government budgets. The success of such partnership depends on the regulation role that government exercises to protect the consumer against monopoly, overpricing, and degradation of service quality. The success of the management contract for the Greater Amman area is attributed to the greater efficiency of the private sector management and adequate funding of the contract, as opposed to WAJ management and shortages in funding. A regulatory framework is a necessary but not sufficient condition for effective regulation by government; rather, it is the effective implementation of the framework that makes the difference.

A clear gap exists between water supply and demand, with increasing shortages foreseen in the future. Cooperation with bilateral and multilateral lending agencies has succeeded in providing about 42% of the funding for projects planned with the year 2010 as a time horizon. Delays in implementation will be at the expense of agricultural water. Water tariff is a highly politically sensitive issue and cannot be left to private firms to decide. Household demand and supply functions are yet to be established. The household sources of water supply should be characterized along with quality of water service, cost of water, and levels of water charges in relation to household income, and policy-related implications should be drawn based on cross-section household survey data.

The high nonrevenue water percentage is testimony to WAJ management inefficiency. Focus on renewal of distribution networks did not substantially reduce this percentage. The upward adjustment of tariffs in 1997 brought about increases in revenues. Law-abiding citizens carried the financial burdens of WAJ inefficiencies at no fault of their own. The upward adjustment of tariffs in 1995 for irrigation and 1997 for WAJ supplies helped increase irrigation water efficiency.

Municipal and industrial water is heavily subsidized by the Treasury. In 2000, the operational costs of water and wastewater, excluding capital costs, amounted to 2.5% of the GDP, of which the consumer paid 1.46% and the government paid the rest. Average cost of municipal water between 1993 and 2002 was the equivalent of US$1.115, compared with a water revenue of US$0.64 over the same period. Costs and revenues of wastewater are not included. Irrigation water is also subsidized, as cost recovery amounted to about US$0.016, compared with a cost of service of US$0.052/m³, or about 30% of total cost in 2002. Further upward adjustment of the price of irrigation water without improving farm income will seriously jeopardize the sustainability of irrigated agriculture in the country.

The overcharges imposed by WAJ as per Regulation 85 for the year 2002 (the Groundwater Bylaw) are likely to reduce overabstraction from groundwater but not eliminate it. The most probable response to the charges for overabstraction will be to increase water use efficiency, adopt automated systems of irrigation to reduce labor cost, and increase the yield per unit flow. Farming will become more intensive, and overabstraction will continue in the Badia.

Pricing and rationing can affect agricultural water use. Analyses of cases in the text show the impact of price adjustment and calculate price elasticity. Upward adjustment will reduce open-field cultivation in favor of plastic houses and thus reduce the diversity of crops. When the World Bank extended the Agricultural Sector Adjustment Loan to Jordan, it was envisaged that upward adjustment of water tariffs would reduce water-intensive crops, but the results were to the contrary. However, increasing the price of irrigation water under a quota system does not produce the desired effect of water savings.

The Water Economic Project model uses shadow prices of water to help decide on infrastructure projects of water transfers. By embodying the economics of water, this tool can provide a powerful aid to water decisionmaking in Jordan. It can be used to evaluate both the systemwide benefits of infrastructure projects and the systemwide costs of different water policies. Finally, it can aid in water negotiations and help guide international cooperation regarding water.

References

Allan, T. 1991. Personal communication between Professor Tony Allan, School of Oriental and African Studies, University of London, and Munther J. Haddadin, as Allan was introducing the concept of "virtual water."

Amir, I., and F. Fisher. 1999. Analyzing Agricultural Demand for Water with an Optimizing Model. *Agricultural Systems* 61: 45–56. Reprinted with minor changes as Chapter 3 of Fisher et al. 2005.

———. 2000. Response of Near-Optimal Agricultural Production to Water Policies. *Agricultural Systems* 64: 115–30. Reprinted with changes as the Appendix to

Chapter 5 of Fisher et al. 2005.

Fisher, F., A. Huber-Lee, I. Amir, S. Arlosoroff, Z. Eckstein, M. Haddadin, S. Hamati, A. Jarrar, A. Jayyousi, U. Shamir, and H. Wesseling. 2005. *Liquid Assets: An Economic Approach for Water Management and Conflict Resolution in the Middle East and Beyond.* Washington, DC: Resources for the Future.

Forward. 1998. *Cost Tariff Model, MWI/JVA.* Amman, Jordan: Ministry of Water and Irrigation.

Haddadin, M. 2003. Significance of Shadow Water to Irrigation Management. Paper sponsored by the University of Oklahoma, Norman, and Oregon State University, Corvallis, submitted to Prince Sultan Bin Abdulaziz International Prize for Water.

———. 2005. Unpublished data in the files of the Directorate of Operation and Maintenance, Deir Alla, Jordan.

Liu J., H. Savenije, and J. Xu. 2003. Water as an Economic Good and Water Tariff Design Comparison between IBT-con and IRT-cap. *Physics and Chemistry of the Earth* 28: 209–17.

Salman, A. 1994. On the Economics of Irrigation Water Use in the Jordan Valley. PhD diss., University of Hohenheim, Germany.

Salman A., E. Al-Karablieh, and F. Fisher. 2001. An Inter-seasonal Agricultural Water Allocation System (SAWAS). *Agricultural Systems* 68: 233–52.

WAJ (Water Authority of Jordan). 2004. Unpublished data in the files of WAJ, Amman, Jordan.

Wolff, H., and A. Nabulsi. 2003. Aspects of Water Resource Management and Hydrosolidarity on the Level of Farming Systems and Households in the Eastern Jordan Valley. Paper presented at German Tropical Day 2003, Technological and Institutional Innovations for Sustainable Rural Development, Goettingen, Oct. 8–10.

World Bank. 1989. *Towards an Agricultural Sector Strategy: Sector Review Report.* Washington, DC: World Bank.

———. 2003. *The Hashemite Kingdom of Jordan Country Assistance Evaluation.* Operations Evaluation Department, Report no. 26875-JO. Washington, DC: World Bank.

Notes

1. This is generally the advice of the World Bank as to managing water.

2. BOT is a system employed in building projects under private sector finance, whereby the builder undertakes to operate the project and collect revenues for an agreed number of years.

3. It is important to understand that demand, in particular, is not independent of price—even in the case of water. If demand and supply are equal only at an unacceptably high price, then water is truly scarce—the price reflecting the scarcity rent—and something must be done.

4. Agricultural needs were calculated at 1,500 m^3/cap/y, industrial needs at 125 m^3/cap/y, and municipal needs at 75 m^3/cap/y.

5. Farmers contest the charges and claim that the JVA's operation and maintenance

service is overstaffed and run inefficiently.

6. There are 1,000 fils to the Jordanian dinar (JD), which is equivalent to US$1.41. Hence, a U.S. cent is equal to 7.08 fils.

7. These delivery charges apply to a standard farm unit of 3.5 ha in area. If the farm unit is larger or smaller than 3.5 ha the upper limit of the consumption block is adjusted linearly accordingly. For example, if a farm unit is 5 ha in area, then the upper limit of the first block for which the first rate of US$0.011 is charged will be $(2,500 \text{ m}^3/3.5)$ = 3,571 m^3/month, and so on for the rest of the consumption levels of the other blocks.

8. It should also be noted that the overabstraction charge is not a charge for the scarcity rent of the water, but a charge to prevent harm to the aquifer involved.

9. One dunum is 1,000 m^2, or one-tenth of a hectare.

10. For a more extended discussion, see the section on capital costs in Chapter 2 of Fisher et al. 2005.

11. This is considerably greater than the reduction predicted by the regression equation.

12. The regression line shown in Figure 6-1 was derived by regressing the quantities generated by the optimizing model on the corresponding prices, not the other way around. However, we have followed the universal practice of showing demand curves with price on the vertical axis and quantity on the horizontal. This means that deviations of the model-generated points from the regression line should be measured horizontally rather than vertically.

13. Models have also been constructed for Israel and Palestine, and preliminary work has been done for models for Lebanon and Saudi Arabia.

14. For the most comprehensive treatment, see Fisher et al. 2005.

15. At present, the WAS model is a single-year model, with the conditions of the year variable by the user. A multiyear model is under construction.

16. So are capital costs, but they are handled outside the actual model in accordance with the discussion above.

17. Capital costs are not to be recovered in the price of the water, but are charged for separately, if not subsidized by the government. See our earlier discussion.

18. In actual runs of the model, the imposition of this third constraint does not appear to be required.

19. It also includes the per-cubic-meter shadow price on the capacity of infrastructure used.

20. The project was later amended and integrated into the Mujib Project. It began in 2003 and starts operation in 2006.

21. It is possible that the aquifer extends much closer to Amman, only 80 km distant, which would considerably reduce the transport cost. This possibility is still under exploration.

22. Note, however, that it also assumes that no other relevant infrastructure will be built. The discussion of the proposed Red Sea–Dead Sea Conduit that follows gives possible implications of this type of consideration.

7

The Role of Trade in Alleviating Water Shortage

Munther J. Haddadin, Amer Salman,

and Emad Karablieh

The supply side of water resources in Jordan as addressed in Chapter 1 indicates their modest availability compared with the levels of demand. A gap exists between water resource supply and demand that evidently is widening with time by virtue of population growth, improved living standards, and the likely loss of certain resources, especially groundwater reservoirs, as a result of environmental degradation or mismanagement. Policies adopted to bridge the gap between supply and demand are fundamental to water resource management.

Jordan's basic water needs include that required for municipal, industrial, agricultural, and more recently, environmental purposes. The aggregate water resources in the country are not sufficient to meet those basic demands. The current flows of water used do not meet the requirements of any of the above user sectors. Jordan's water picture is not unique, but very likely similar to the water situations in many other semiarid countries. The determination of the gap between the supply and demand for basic needs requires the calculation of both sides of this classic supply–demand equation.

The sources of blue water are recorded, and the records are kept at the Ministry of Water and Irrigation (MWI); gray-water sources are also measured and updated. Desalinated water (silver water) flows are known, and records are kept at the MWI. But green water is neither attended to by the MWI nor measured by any authority, even though it constitutes a substantial component of the renewable water resources. Green water's share of official attention has been qualitative, such as describing the season as good when it is rainy and lean when there is a drought. But farmers know as much without the government officials' appraisal of the season.

In this chapter, the constituent factors of supply and demand are assessed, with due consideration to green water. A method is introduced to quantify green water in Jordan, enabling the determination of the imbalance in the population–water resources equation. The analysis proceeds to determine the stress to which the water resources are subject and the strains that result from that. Stress–strain relationships are derived, and a water modulus is defined as the rate of change of water stress with respect to strain.

The concept of shadow water is introduced, essentially reflecting the water savings a country realizes by importing commodities. Shadow water can bridge the gap between supply and demand. A rigorous analysis is made to assess the cost of shadow water and compare it to the real cost of indigenous water. The trade in food commodities is reviewed, and the resulting water trade is analyzed. This is the first such attempt done for Jordan.

Supply Side of the Population–Water Resources Equation

On the supply side, Jordan's natural renewable water resources are blue water, consisting of surface water and groundwater, and green water, or soil water, which is stored around soil particles and in soil pores by surface tension and capillary action (Chapter 1). Soil water supports rain-fed agriculture, including rangeland, as plant roots are the only users capable of extracting this type of water. Other renewable water resources in Jordan are created by humans and consist of treated wastewater, or gray water, and desalinated brackish water, referred to here as silver water. Seawater desalination, although widespread in the other Gulf states, has not yet started in Jordan. In addition, the country and its neighbors have nonrenewable water resources consisting of stocks of fossil water that accumulated in aquifers thousands of years ago. When a reasonable balance exists between supply and demand, use of such nonrenewable stocks is not a rational policy to follow on environmental grounds. Rather, it is recommended that such resources be set aside as a strategic reserve. Another form of nonrenewable water supply is the quantities abstracted from renewable groundwater aquifers over and above their sustainable yields.

The components of the supply side of the equation, with the exception of the important component of soil water, can be quantified using technological instruments such as flow measurement devices. Safe yields of renewable aquifers can be calculated too. Most water authorities focus on the assessment of these easily measurable resources, but few of them ventured to assess the availability of soil (green) water and to express it in terms of irrigation (blue) water equivalent.

The average surface-water resources of Jordan total 885.6 mcm/y (Table 1-1), of which about 840 mcm can be economically mobilized; the groundwater resources average 266 mcm/y (Table 1-2); the desalinated (silver) water averaged 8.76 mcm/y in 2003 and about 45 mcm/y in 2005 (MWI 2004); the treated wastewater effluent averaged 80 mcm in 2002 (Chapter 4); and the leakage from networks and irrigation return flows averaged 70 mcm that year. The soil (green) water as a supply component remains to be calculated (Haddadin 2002; Haddadin 2003a).

Blue-Water Equivalent of Green Water

The basis of calculating the blue-water equivalent of green water is the equivalence in plant yield per unit land area of green water and the parallel yield of blue water used in irrigation. The yield per unit land area of rain-fed agriculture varies with the frequency and intensity of precipitation during any given season, the soil type, and other important factors relating to relative humidity, evaporation rates, farming methods, and technology. Agricultural inputs such as type of seeds, fertilizers, machinery, farmers' skills, and capital intensiveness are factors as well. In most cases, at least in arid and semiarid countries, the productivity per unit land area of rain-fed agriculture does not match the corresponding productivity of irrigated areas, for several reasons: the controlled water application over irrigated areas as opposed to the irregular precipitation over rain-fed areas; the advanced technology employed in irrigated agriculture; the capital-intensive farming methods; the focus of research and extension activities on irrigated agriculture; and higher cropping intensities in irrigated areas.

Worldwide, the irrigated areas in 1998 constituted 18% of the cropped areas and produced 40% of the world's food (Gleick 2000). Postel 1999 quotes the 1996 FAO Production Yearbook, reporting that irrigated land constituted 17% of the cropland and accounted for 40% of the food production in the world that year. The average of the two reports indicates that 17.5% of cropland was irrigated and produced 40% of the world's food. The remaining 60% of food production can thus be attributed to the rain-fed cropland, whose area average was 82.5%. From the above, two simultaneous equations can be derived:

$$0.175 \times A \times p_i = 0.4\ Y \text{ and}$$
$$0.825 \times A \times p_r = 0.6\ Y$$

where
 A is the total cropped area in the world,
 Y is the amount of food produced in the world,
 p_i is the productivity per unit land area of irrigated agriculture, and
 p_r is the productivity per unit land area of rain-fed agriculture.

Solving the above equation yields

$$p_i/p_r = 3.1428.$$

This quotient, 3.1428, happens to be, by sheer coincidence, nearly identical to the value of π (pi), the constant ratio of the circumference of a circle to its diameter, which is equal to 3.141592 (Johnson 2003). A similar ratio was obtained for the productivity of irrigated wheat crops in Jordan and the wheat crop yield of rain-fed areas on the Jordanian Plateau with a rainfall average of 350 mm (MOA 1975). Although a rational linkage between the productivities' ratio and the universal constant π does not seem to exist, it is convenient to remember.

The flow of water, at its source, needed to produce a certain intensity of soil moisture in a unit area of irrigated land is greater than the precipitation amount that produces the same intensity in a unit area of rain-fed lands. The amount of irrigation water, measured at the source, is reduced throughout the conveyance, distribution, and application stages before it gets to the soil. This is where irrigation efficiency makes a difference.

A weighted average for irrigation efficiency for the Middle East, with its countries belonging to the high, upper-middle, lower-middle, and low income categories, was calculated to be 0.60 (Haddadin 2002; Haddadin 2003a, 32).[1] Thus to produce a cumulative soil moisture of 375 mm, which is the precipitation needed for a meaningful crop yield on rain-fed land in Jordan, the equivalent irrigation water flow at the source that produces an equivalent soil moisture should be (375/0.6 = 625 mm).

Because the irrigation yield is π times the rain-fed yield of the unit land area, the irrigation-water equivalent that produces the equivalent rain-fed yield is the above irrigation depth of 625 mm divided by π, or

$$W_{soil} = 625 \text{ mm}/\pi$$
$$= 200 \text{ mm}.$$

The soil moisture produced by the 375 mm of rainfall in Jordan is worth the application of 200 mm of irrigation water measured at the source. The irrigation (blue) water measured at the source that has the effect of soil moisture generated by the meaningful rainfall in Jordan, is therefore of 200 mm depth. In a similar approach, the quantity of green water that can support rangeland in Jordan was calculated to be equal to 25 mm of irrigation-water depth (Haddadin 2002, 2003a).

The contribution of soil water to agricultural water resources in terms of irrigation-water equivalent is therefore expressed by the following equation:

$$W_{soil} = 200 R + 25 r,$$

where
   R is the total area of rain-fed agricultural land, in dunums (du), and
   r is the total area of rangeland, in dunums.[2]

Applying this equation to Jordan (3.13 million du of rain-fed area and 7.91 million du of rangeland), the blue-water equivalent to Jordan's green-water potential (cropland areas plus rangeland) for the year 2002 is calculated as follows:

$$W_{soil} = 200 \times (3.138192) + 25 \times (7.910)$$
$$= 866 \text{ mcm.}$$

On an overall basis, the green water has the potential to provide the equivalent of 866 mcm of blue water. For a population of 5.329 million people in 2002, the per capita share of green water was 155 m^3. Obviously, the contribution per capita of green water diminishes with time as population grows and as good rain-fed lands are lost to urbanization and rangelands to erosion and mismanagement.

Total Renewable Water Resources

The natural renewable freshwater resources of Jordan are stated in Chapter 1 as

$$\text{renewable freshwater potential} = B_w + G_w + I_W,$$

where
   B_w is the indigenous blue water (752 mm^3/y),
   G_w is the indigenous green water (866 mm^3/y), and
   I_W is the transboundary blue water (352 mm^3/y).

Solving the above equation,

$$\text{renewable freshwater potential} = 1,970 \text{ mm}^3/y.$$

The future renewable water potential will increase with greater contributions from silver water and municipal wastewater treatment. The increase is dependent on investment in generating silver water from brackish-water desalination and reclaiming municipal wastewater. Investment in the second entails expanding wastewater collection networks, increasing the capacities and upgrading the methods of wastewater treatment facilities, and developing human resources.

In 2002, the total resources potential consisted of 2,120 mcm with the addition of about 10 mcm of silver water and gray water. The per capita share that year thus amounted to 396 m^3.

Demand Side of the Population–Water Resources Equation

The demand side includes the water required to satisfy the basic needs of municipal, industrial, agricultural, and environmental uses. The basic environmental needs are location specific and hard to quantify and thus will not be addressed here.

The demand for water is primarily a function of the standard of living of consumers. The income level of a household, in addition to social and cultural factors, determines its consumption habits and limitations. As a guide, the analysis herein adopts the income level of a country for estimates of the demand for water, using the World Bank's income categories, as listed above, to determine the level of water demand for each income group. Additionally, a distinction must be made between actual consumption and the desired level of demand. In Jordan, for example, actual consumption is lower than the demand level because of the higher tariff charged to higher brackets of water consumption, the rationing of municipal water and irrigation supplies, and the rotation system used for service. Water pumped in municipal networks averaged 47 m^3/cap in 2002. The suppressed supply in the municipal networks impacts the demand of industries that depend on those networks for their water supply.

Water Thresholds in a Virtual Environment

A water threshold, as used herein, is the quantity of water per capita needed to meet each of the basic needs for municipal, industrial, and agricultural uses. Each has its threshold indicating the demand level that is desired to meet its requirement.

The production environments in the countries of the world are hardly suited for each country to produce all its needs of goods. In fact, this is simply undesirable in economic terms, with benefits arising from the division of labor, absolute and comparative advantages, and other factors. Water is an essential input component in the production of goods, and therefore the import of goods results in savings in water consumption, in addition to other resources. To evaluate the water savings that imports make possible, a virtual environment is envisioned in which a country, against sound economic policy in real life, is enabled to produce all its needs of goods. In a virtual environment, all conditions needed for the production of goods are assumed to be within the reach of each country. A country in the tropics, for example, can produce needed goods that are impossible to produce in its natural climatic zone. Similarly, a developing country is enabled to produce the industrial commodities needed by its population, when in real life, the market size, technological qualifications, human resources, raw material, and capital outlays stand as barriers in the way of such production. Although the actual production of all commodities within a small market like Jordan is undesirable in economic terms, the virtual environment concept helps in evaluating the indigenous water flow that would be needed to produce the imported goods.

Water flows needed to produce the industrial goods, called the industrial water threshold, are calculated, and so are the flows needed to grow the food commodities, or the agricultural water threshold. Because soil water (green water), basic for the production of rain-fed agriculture and natural rangelands, is counted on the water supply side as outlined above, the agricultural water threshold in a virtual environment is therefore based on zero residual soil moisture, and the water resources, including green water, are called upon to supply the needed amounts.

The water threshold is calculated at the source of supply so that the supply and demand quantities are compared at the same location. This necessitates the assumption of certain values to be assigned to the efficiency of water conveyance and distribution, as indicated below.

Shadow Water

Shadow water is a term introduced here to signify the amounts of indigenous water saved by commodity imports. It is similar to the notion of virtual water but is location and quantity specific and covers more than just agricultural imports. The term *virtual water* has been in use in the literature since it was first coined in 1996 to address the impact of food trade on water resources of water-strained countries like those of the Middle East.[3] That term is used primarily in the context of foreign trade in food commodities and is not location specific, whereas shadow water includes all imported commodities and is location specific in the sense that it pertains to the country that imports commodities (both food and industrial) from the outside. Shadow water is defined as the indigenous water that would be needed to produce the imported commodities in the importing country.

Because production water has to be real and cannot be virtual, the intended meaning of the term *virtual water* is best reflected, in our judgment, by the term *virtual flow* of water that the import of the commodities entails. The imports entail a virtual flow of all the inputs of production not visible in the imported commodity, not just the water input. In food commodities, for example, the virtual flow includes seeds, water, fertilizers, pesticides, labor, machinery, and even arable soil, which are all production inputs. This is not meant to debate the use of the term *virtual water*, however, but to justify the use of the new term *shadow water*.

In the dictionary, one definition of *shadow* is "having form without substance (~ government)."[4] In economics, the term *shadow prices* is used to denote not the ruling market prices, but prices that reflect social costs and benefits. The amount of indigenous water needed to produce the imported commodities in the importing country has, in that sense, a form (the traded commodity) but no substance (no actual water flows from the exporting to the importing country).

The virtue of location specificity that shadow water has is clear: the amount of water needed to produce, say, a ton of wheat in the exporting country could be different from the amount of water that would have been used to produce the same ton of wheat in the importing country. In fact, the quantity of water that would have been used in the importing country to produce the imported commodities (but was saved by the imports) is the *shadow* of the quantity of water actually used in the exporting countries to produce those commodities. The shadow is not necessarily equal in size to the object it reflects. Concurrently, one cannot physically catch or feel the water saved by imports, nor can one get his or her hands wet with virtual water "embedded" in an imported commodity, in just the same way that a person could not catch or feel a shadow. In a market economy, the shadow price not only is a reflection of the economic value of a certain commodity, but also embeds in it its social value; likewise, shadow water has social costs and benefits, as shall be discussed later. In politics, a shadow cabinet does not have any executive powers that the real cabinet commands; likewise, shadow water does not have the "power" that real water commands, gaining power only after the commodity reaches the markets of the importing country.

This concept of shadow water thus embraces the concept of virtual water. The term *virtual water* was later contested by Stephen Merett (2003), who rightfully argues that it does not reflect the description of what it is intended to mean. The new term *shadow water* resolves the difference between Allan and Merett without departing from the intended meaning.

Municipal Water Threshold

We have made calculations of the annual withdrawal for consumption of municipal water in the four income categories of the world, using a recent database (Gleick et al. 2002). The world annual averages are 137 m³/cap/y for the high income category, 65 for the upper-middle, 41 for the lower-middle, and 20 for the low, measured at the water source (Gleick 2004). The corresponding averages for these categories in the countries of the Middle East, using the same database, are 125 m³/cap/y for the high income category, 71 for the upper-middle, 48 for the lower-middle, and 11 for the low. Guided by the above averages, which reflect the municipal water requirement at the point of use and not at the source, and by his experience in the region, Dr. Munther Haddadin has adopted the consumption rates of 90 m³/cap/y for the high income category, 73 for the upper-middle, 60 for the lower-middle, and 38.5 for the low. At the water supply source, the quantities are magnified to account for the conveyance and distribution efficiency. In this regard, the efficiency a country can achieve is a function of its development category. Richer countries can achieve higher efficiencies than poor countries. The efficiency achievable by the countries of the four income categories is assumed to be 0.9 for the high income category, 0.85 for the upper-middle, 0.8 for the lower-middle, and 0.7 for the low. The corresponding municipal water threshold at the source of water for these income categories would be 100 m^3/cap/y (up from 90) for the high income category, 85 (up from 73) for the upper-middle, 75 (up from 60) for the lower-middle, and 55 (up from 38.5 m^3/cap/y) for the low. Being a lower-middle income country, Jordan would thus have a municipal water threshold of 75 m^3/cap/y as compared with an actual average of 47 m^3/cap/y served in 2002, a shortage of 28 m^3/cap/y. This shortage creates a municipal *water strain,* which a society sustains resulting from *water stress,* expressed as the number of persons in the country per unit flow of its water resources.[5]

Industrial Water Threshold

The industrial water threshold per capita is the amount of water needed to produce the commodities a person in a given country needs per year under virtual environment conditions. Using the above-referenced database, the industrial water withdrawal in the world for the year 2002 was 705,496 mcm. The population of the world was 6,104.8 million persons. The industrial water at the point of use was calculated to be 603,220 mcm.[6] Using the efficiencies assumed for the municipal networks, the average industrial water allocation at the source per person was calculated to be 98.81 m^3 that year.

However, to estimate the industrial water use per capita in each income category, an assumption has to be made in absence of better data. It is assumed herein that the water ratios for the four income categories at the point of use in a virtual environment are in the order of 5:3:2:1, respectively; in other words, the industrial water needed in the high income category is five times that in the low income category. This discrimination follows the affordability to use industrial goods in the four income categories. At the point of use, the water amounts would be 235 m^3/cap/y in the high income category, 141 in the upper-middle, 94 in the lower-middle, and 47 in the low. At the source, these amounts become the thresholds of 260 m^3/cap/y in the high income category, 165 in the upper-middle, 120 in the lower-middle, and 65 in the low. Jordan, in a lower-middle income category, would have an industrial water threshold of 120 m^3/cap/y compared with about 7 m^3/cap supplied in 2002, a shortage of 113 m^3/cap that year covered through commodity imports, such as vehicles, machinery, and refrigerators.

Agricultural Water Threshold

We also conducted an analysis of diets in the Middle Eastern countries, using the database supplied by the Food and Agriculture Organization of the United Nations, and examined the average diet of the industrialized Western countries for comparison. The average caloric intake is shown in Table 7-1.

TABLE 7-1. Reported Average Daily Caloric Intake per Capita

Source of calories	Low income	Lower-middle income	Upper-middle income	High income	Industrialized Western countries
Plant-based	1,905	2,711	2,561	2,615	2,462
Animal-based	136	300	433	716	1,049
Total	2,041	3,011	2,994	3,331	3,511

Source: FAO 2000.

TABLE 7-2. Generalized Caloric Intake per Capita per Day

Source of calories	Low income	Lower-middle income	Upper-middle income	High income
Plant-based	1,950	2,450	2,650	2,650
Animal-based	250	350	450	650
Total	2,200	2,800	3,100	3,300

Source: Author's estimate guided by analysis of results in Table 7-1.

TABLE 7-3. Agricultural Water Requirements to Produce Generalized Caloric Intake (m^3)

Source	Low income	Lower-middle income	Upper-middle income	High income
Plant-based	393	494	534	533
Animal-based	287	400	514	743
Total	680	894	1,048	1,276

Source: Table 7-2, obtained by multiplying plant-based calories by 0.201 and animal-based calories by 1.143.

Guided by Table 7-1, we formulated a generalized caloric intake, as shown in Table 7-2. A food diet table is used for the lower-middle income category of countries in the Middle East (Elmusa 1997). The amount of water needed to produce the food of a diet composed of 2,400 plant-based and 300 animal-based calories was found to be 827 m^3/y (484 m^3 for plant-based calories plus 343 m^3 for animal-based).

The water requirement to produce a plant-based calorie is thus 0.201 m^3 and for an animal-based calorie, 1.143 m^3. Applying these requirements to the generalized caloric intake table, one obtains the water requirements to produce the calories needed for each income category, before adjustment for plant productivity, as shown in Table 7-3.

An examination of productivity reveals that the advanced industrial countries have an overall productivity per unit land area equal to 140% of the

TABLE 7-4. Jordan's Water Threshold, Water Supply, and Shadow Water, 2002 (m³/cap/y)

Water	Municipal	Industrial	Agricultural	Total
Threshold (demand)	75	125	1,500	1,700
Resources (supply)	47	7	252[a]	306
Balance	(28)	(118)	(1,248)	(1,394)

a. Includes 49 m³/cap of nonrenewable groundwater, 14 m³/cap of wastewater, and 155 m³/cap of soil water (green water). The sustainable portion of agricultural blue water was 34 m³/cap.

lower-middle income countries of the Middle East (FAO 2004). This difference was achieved using 0.7 times the amount of water the Middle Eastern countries used. The productivity of the high income category per unit water flow is thus equal to 200% of that achieved by the lower-middle income.

An adjustment of Table 7-3 is therefore warranted to account for the productivity of calories per unit flow of water. The lower-middle income base is assumed equal to 1.0. With that, the high income would be 2.0 (200% productivity ratio). The upper-middle is interpolated to be 1.3 and the low income category is extrapolated to be 0.7. With these productivity weights, the water requirement, measured in m³, to produce the above calories would amount to 971 for the low income category, 894 for the lower-middle, 807 for the upper-middle, and 639 for the high.

A final adjustment in the agricultural water requirement is needed to account for the efficiency of conveyance and distribution of irrigation water. The efficiencies are assumed at 0.70 for the high income category, 0.65 for the upper-middle, 0.60 for the lower-middle, and 0.55 for the low. Applying these efficiencies to the above water requirements, one obtains the water threshold at the source of 912 m³/cap/y for the high income category, 1,241 for the upper-middle, 1,490 for the lower-middle, and 1,765 for the low income category. If these amounts are rounded, they could be set at 900 m³/cap/y, 1,250, 1,500, and 1,780, respectively.

Total Water Threshold and Water Equilibrium

The summation of the water thresholds calculated above, in a virtual environment, constitutes the total water threshold for municipal, industrial, and agricultural purposes. These totals amount to 1,200 m³/cap/y for the high income category, 1,500 for the upper-middle, 1,700 for the lower-middle, and 1,900 for the low. Each comprises the demand side of the water equation per capita for the particular category. The comparison with the amounts supplied for each purpose enables the determination of the deficit or surplus that a country has in a given year. Table 7-4 shows the case for Jordan in 2002.

The shadow water closes the deficit between the water threshold (demand) and the water supply for agricultural and industrial uses. It does not help meet municipal demands. Only real water imports can help in this regard, such as has been the case so far in importing bottled water for drinking purposes, or as the case may be in the future if regional water projects are implemented to transfer municipal water from outside resources. Without real water imports, the gap between supply and threshold (desired demand) translates into a shortage of municipal water, which generates strains in society. This situation may also necessitate the diversion of blue water used in agriculture to municipal uses. This was done in Jordan, in cases mentioned in Chapters 2 and 4, and in Damascus, Syria, when Ein Al Fiejeh, feeding the historic Barada River, for which Damascus was famous, was diverted to municipal uses.

In checking the supply–demand equation, it is important to include, on the supply side, all water resources and their substitutes: blue and desalinated water for municipal purposes; blue and shadow water for industrial purposes; blue, soil, gray, and desalinated water where applicable; and shadow water for agricultural purposes. The equilibrium of each component should be checked separately. A check can be made on all the supplies and the sum of the thresholds to examine the overall water equilibrium in the country.

Globalization of Water Resources

In addition to its crucial role in maintaining water equilibrium in the importing countries, shadow water reflects the role that foreign trade plays in water resource management of all countries of the world. Countries with surpluses in commodity trade are net exporters of water, and countries with deficits in that trade are net importers of water. In a sense, the free transfer of goods across borders of sovereign nations is an act reflecting the globalization of water resources (Haddadin 2003b). It is a process by which water-poor countries can call upon the water resources of other countries to bridge the gap between supply and demand in the industrial and agricultural sectors. Even hydropower imports have been promoted to help meet energy demands in energy-importing countries.[7]

The role of trade in alleviating water shortage is thus obvious. It is the reason why extremely water-short countries may exhibit no serious sociopolitical problems. In Jordan, although indigenous water supply in 2002 (306 m³/cap) amounted to only 18% of the aggregate water threshold of 1,700 m³/cap (Table 7-4), one does not hear loud screaming or witness massive protests; shadow water covered about 80% of the aggregate water threshold, leaving only 2% of the total threshold unprovided, and that was reflected in municipal water shortages that have been raising complaints and causing inconvenience to consumers almost every summer.

Free trade is very helpful to Jordan and all other water-short countries. The government policy in this regard has been clear in seeking to become a member of the World Trade Organization (WTO), obtaining the privilege of trade partnership with the European Union, earning free-trade status with the United States, and promoting qualified industrial zones, whose production enters the U.S. market exempt from duty taxes. The free-trade policy adversely impacted the agricultural sector, as it had to compete on the domestic market with imported commodities. A net importer of commodities is primarily a water-strained country, and its trade policy cannot be but a reflection of its dependence on other countries for survival. Water, in this regard, affects not only domestic policies of social and economic dimensions, but also foreign policy in many ways.

Water Stress and Water Strain

The term *water stress,* widely used in the water literature, is not properly anchored in basic sciences as it should be. *Stress* as used in engineering mechanics signifies pressure, or a load (kg) per unit area (cm^2) of the section of a material to which that load is applied. A parallel use of the term in the water field should also signify pressure, in this case the pressure caused by demand for water, which in turn is generated by the population. A proper definition of water stress would thus be the number of people per unit flow of water resources of the country. A unit flow has been known as 1 mcm/y. Therefore, water stress has the unit of persons per unit water flow, or persons per million cubic meters per year. As such, one should avoid the use of such expressions as "the country is water stressed," when in fact, it is the water and not the country that is stressed. Water stress can be expressed in equation form as

$$\sigma = P/W_R$$

where
σ is the water stress, in persons per unit water flow (1 mcm/y),
P is the population of the country, in persons, and
W_R is the potential for indigenous water supply, in million cubic meters per year.

The water stress at equilibrium, σ_0, does not produce any strain. It corresponds to the state when the resources match the aggregate threshold, or

$$\sigma_0 = P/W_{TH}$$

The term *strain* is also used in engineering mechanics, where it is defined as the ratio of deformation that a material under stress undergoes with respect to its original size before the deformation is induced. Stress and strain are cause and effect, and in elastic materials, the rate of change of stress with respect to strain is called the modulus of elasticity. To reflect that in the water literature, it is proper to say that a country is water strained but not water stressed. When the water resources match the aggregate threshold, the water stress is that of equilibrium, and the water strain vanishes.

The water strain is measured as the ratio between the water shortage or shadow water (the deformation) and the water threshold at which the society or country exercises zero strain. Water strain can be expressed in equation form as

$$\varepsilon = W_{SH}/W_{TH}$$

where

ε is the water strain exercised by the society, dimensionless,

W_{SH} is the aggregate shortage (shadow water), in cubic meters per capita, and

W_{TH} is the aggregate water threshold, in cubic meters per capita.

The *water modulus*, measured in units of water stress (strain is dimensionless), is the slope of the stress–strain diagram, or the rate of change of stress with respect to strain:

$$E_W = d\sigma/d\varepsilon$$

where

E_W is the water modulus, and

$d\sigma/d\varepsilon$ is the rate of change of water stress with respect to water strain, or the slope of the water stress–strain diagram.

The higher the water modulus, the higher the strain that society feels, and vice versa. Figure 7-1 shows a typical stress–strain curve and the water modulus. The intersection of the curve with the stress axis is the stress at equilibrium, and that value varies with the income category of the country. The equilibrium stress is 833 for the high income category, 667 for the upper-middle, 588 for the lower-middle, and 526 for the low. The shape of the curve for each is determined by the equation

$$\sigma = 10^3/[W_{TH} (1.0 - \varepsilon)]$$

where σ, ε, and W_{TH} are defined as above.

When ε vanishes, the value of σ becomes the equilibrium stress, σ_0, whose values for the four income categories are as stated above. Figure 7-2 shows the characteristic curve for each income category.

The above definitions and mathematical formulations enable the drawing of

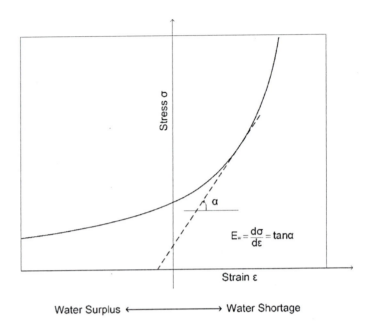

FIGURE 7-1. Typical Stress–Strain Diagram and Water Modulus

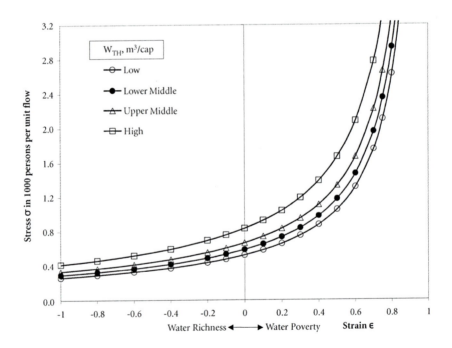

FIGURE 7-2. Characteristic Curves for Each Income Category

water characteristic curves (stress–strain curves) for countries or group of countries belonging to the same income category. A characteristic curve could be drawn for the world after determination of its water threshold and potential of water supplies. Such a curve would show the countries of surplus and the countries of deficit and would clarify the role that international free trade could play in the alleviation of water shortages. An attempt to do that will not be made here, however.

Advantages and Drawbacks of Shadow-Water Relief

Among the many virtues of shadow water, the indigenous reflection of exogenous water used to produce the imported commodities is the relief it provides from water stress in arid and semiarid countries, where water stress and societal strains are felt to varying degrees. Without such relief, the insufficient indigenous water resources would have to be stretched to meet at least the very minimum demand levels by society for some import substitution. The result would be high water stress and substantial social and environmental strains in society. The relief that shadow water provides through commodity imports, a practice known for millennia, has several positive impacts in addition to the financial advantage over the indigenous marginal cost of water in the importing country. Obviously, in the absence of commodity imports, shadow water is nonexistent; without exogenous water needed to produce imported commodities, there can be no shadow. The following advantages (benefits) are associated with shadow water:

- Shadow water relieves water stress in the importing country because it provides a replacement for agricultural and industrial water that otherwise would have to be allocated in the importing country. Current agricultural uses in many semi-arid countries account for a high percentage of the consumption of water resources (68% in Jordan to 90% in Egypt, for example). Replacement of such allocation percentages, or part thereof, undoubtedly frees substantial amounts of indigenous water resources that can be diverted to other beneficial uses (municipal, industrial, and environmental).
- It shields the importing country from the negative consequences of water strains on society as demand for water outstrips supply. Such strains could trigger social unrest, with political ramifications that could spill over to neighboring countries through waves of "environmental refugees."
- It averts several negative impacts that trigger economic consequences and losses. A reputation of water stress and associated strains is hardly conducive to the flourishing of the tourism market or the promotion of exports, all of which are water dependent to one degree or another. Tourism, considered an export services market, and the other export commodities would undoubtedly suffer setbacks that would trigger a chain reaction affecting

employment, family incomes, and standard of living. This could fuel anger, social unrest, and protests.

- It reduces to a great extent the burdens that otherwise would have to be carried by the poor. When water stress is high, water prices go up, and accessibility to water by the poor is jeopardized.
- It also greatly reduces the incidence of public health hazards that result from high water stress. The susceptibility of water pollution increases with higher water stress, and so does disease incidence, especially waterborne diseases, which usually push child morbidity higher.
- It provides a grace period to politicians where riparian conflicts over international watercourses exist, deferring the moment of decision regarding riparian issues that would otherwise be more pressing.
- It provides a conduit for the globalization of water resources to the benefit of water-strained economies or societies. Through import of commodities, the importing country shares with the exporting country the benefit of water resources used by the latter.
- Even in countries with a water surplus, shadow water enables the importing country to avert the negative environmental, social, and political impacts of water resource development, such as the construction of dams and water reservoirs.

The drawbacks (costs) of shadow-water relief can be listed as follows:

- The importing country forgoes the economic added value it would have gained had the products been produced domestically.
- The benefits from indigenous production of commodities, especially agricultural ones, are not realized, including the generation of jobs at much less capital investment per job than alternative employment—in industry, for example—would provide. Agriculture helps maintain a balanced distribution of population over the territories of the country and provides means of livelihood for people who otherwise would migrate to the urban areas, with all the social burdens society has to shoulder as a result of such migration. The cost of servicing their social needs in education, health care, housing, and the like would be much higher. The social cost of the retreat of agriculture undoubtedly taxes society.
- Shadow water does not provide food security, especially strategic food staples such as grain and dairy products. Some countries fear the emergence of monopolies in the export of food and other needed commodities; others fear the possibility of trade boycotts. Under normal conditions of free trade, food security cannot mean domestic production of all needed food, but is attained by the ability to provide needed food through domestic production or trade or both. A country is secure when its balance of foreign trade is held in equilibrium.
- Shadow water drains foreign exchange reserves of the country whose shadow

water is substantial. Unless sources are found to generate foreign currency, the country will face difficulties in the stability of its currency and in financing its foreign trade.

The above intangible benefits and costs, substantial as they are, are inherent in shadow water in much the same way as the intangible social costs and benefits are part of the shadow price concept in economic analysis, thereby reinforcing the justifications of the use of the term.

Agriculture in Jordan's Economy

The contribution of agriculture to Jordan's gross domestic product (GDP) at current prices decreased from US$265 million in 1990 to US$193 million in 2002. The share of agriculture in total GDP decreased in parallel, from 7.9% to 2.39%. The declining share of agriculture is a result of structural changes in Jordan's economy; the erratic pattern of rainfall, with droughts in certain years; a decrease in export earnings; and the financial and fiscal crisis the country confronted in 1988. The decline was aggravated in 1994, when, in its measures for economic structural adjustment, the government adopted a trade liberalization policy that was also a precondition to accession to the WTO. As a result, protection of the indigenous agricultural products through import barriers ceased, and the direct subsidy to farmers and animal growers was tapered (Hjort et al. 1998).

The Role of Agribusiness

The fast-growing food and beverage processing industry has increased markedly in Jordan since the 1980s. Increasing populations, rising incomes, changes in lifestyles, and ambitious government development plans contributed to the industry's demand-led growth. By 2000, the food processing industry was contributing an annual sum of US$778 million to the GDP of Jordan. The agroindustrial sector is characterized by a large number of small enterprises (3,178 establishments), which employ more than 15% of the total workforce (23,619 employees). The sector consists mainly of edible oil and fat processing, dairy production, and meat canning. Increased agricultural output since the early 1990s has stimulated more fruit and vegetable canning and juice, beverage, and oil processing, relying mostly on indigenous production. Some industries rely partly or entirely on imported raw material, such as livestock feed, vegetable oil other than olive, and confectioner's sugar. Agricultural output, especially of olives, fruits, and vegetables, increased substantially from 1985 to 2000.

Agricultural Trade

The value of agricultural imports has shown fluctuation over the last decade; the minimum was US$732 million in 1994, and the maximum was US$1,115 million in 1996. Agriculture's contribution to total national imports dropped from 21% in 1994 to about 17% in 2002. The corresponding dollar values of the agricultural imports, however, increased from US$732 million in 1994 to US$861 million in 2002.

Agricultural exports, primarily horticultural products, contribute 10% to the merchandise export trade. The value of agricultural imports was more than 250% of agricultural exports and about 446% of the agricultural GDP in 2002. On the other hand, agriculture's contribution to total national exports has dropped from 20% in the period 1992 to 1996 to about 15% in the period 2000 to 2002.

Most agroindustries were justified on the basis of import substitution. In 2002, their exports amounted to US$171 million and their imports US$353 million. Export markets were primarily the Gulf states, with a fraction directed to Europe, the United States, and other markets. The sector has potentials for expansion if marketing services are improved, investments are promoted, and quality is matched to the standards demanded by the trading partners.

Food Security

Indigenous agricultural production has not been capable of meeting the increasing growth in food consumption, however, induced by rapid population growth and an improved standard of living. The gap between indigenous agricultural production and the demand for food had to be bridged by imports. Agricultural shadow water amounted to 1,248 m^3/cap in 2002 out of some 1,500 m^3/cap needed to close the gap (Table 7-4). It is evident that Jordan's food security depends heavily on food imports, a fact that impacts its water policy as well as its trade and foreign policy.

The Importance of Shadow Water to Jordanian Life

The above shows in no ambiguous terms that Jordan has been relying on agricultural imports and sustaining deficits in its foreign trade for decades. The volume of water available for food and agricultural production has not been sufficient to meet increasing demands for irrigation water. Higher incomes and improving living standards combined with steadily increasing populations and waves of refugees generated substantial increases in food demand that could be satisfied only by increasing food imports. The average annual rate of growth in food imports was 11% in 1964–1974, 14% in 1975–1984, 8% in 1985–1994, and 2.5% in 1995–2003. This means that the exogenous virtual flow of water, termed agricultural shadow water hereinafter, into the country has been increasing each year. Such an increase was reflected in a decline in food self-sufficiency ratios. The increase of import prices, mainly in foodstuffs and especially grains, had a direct effect on the upward trend inflation took in 1996.

Jordan's Agricultural Water Trade

An attempt is made in this section to quantify the agricultural water flows out of the country and the virtual flows through agricultural imports (shadow water) from 1994 to 2002. Sources of data used in the analysis are the Department of Statistics of Jordan (DOS) for export and import quantities; studies sponsored by Gesellschaft für Technische Zusammenarbeit (GTZ) of the Federal Republic of Germany and included by the World Bank (1997) in its report for average crop yields; and an agricultural statistical survey conducted for the same period and cited by the same report for the weighted average of crop-water requirements in the irrigated areas of Jordan. Other sources of information are also referenced below.

Crop-Water Requirements

Various methods have been developed to determine the water requirements for specific plants. A comprehensive guide to the details of these methods is Doorenbos and Pruitt (1992). The calculation method is not explained here. For more details on the calculation method, consult an authoritative reference such as Critchley and Siegert (1991), Doorenbos and Pruitt (1992), or Allen et al. (1998).

Shatanawi et al. (1998) used the literature above to determine the net water requirements of crops planted in the Jordan Valley according to agroclimatic zones. The crop net water requirements stated below were adopted from their work. To convert the net water requirements to gross water requirements, the irrigation efficiency of the JVA system, reported by the JVA as 72%, is adopted. Using data from these sources, the monthly gross crop-water requirements have been assigned to each crop from different agroclimatic zones in the Jordan Valley. The average irrigation crop-water requirements for main crops produced in Jordan are shown in Table 7-5.

We assessed the use of indigenous water resources over the period 1994 to 2002. The areas of rangeland in the country average 7.91 million du (FAO 1999), and as per the above presentation for the quantification of green water, the contribution of rangeland to green-water resources would amount to 197 mcm.

The areas of field crops, vegetables, and fruit trees of various types planted in the period 1994 to 2002 were multiplied by the corresponding crop-water requirements in Table 7-5 to estimate the total water used to produce these crops in Jordan. This includes blue, green, and gray water, but does not include

TABLE 7-5. Average Irrigation Crop-Water Requirements (CWR) for Crops Produced in Jordan (m³/ha)

Crop	CWR	Crop	CWR
Wheat	3,530	Okra	3,790
Barley	2,360	Lettuce	2,070
Lentils	3,500	Sweet melon	3,560
Vetch	2,500	Watermelon	3,560
Chickpeas	3,500	Spinach	2,080
Maize	7,230	Onion, green	5,320
Sorghum	6,000	Onion, dry	8,230
Broom millet	6,000	Snake cucumber	2,480
Tobacco, local	3,000	Turnip	2,480
Tobacco, red	3,200	Carrot	2,370
Garlic	5,230	Parsley	2,450
Vetch, common	3,200	Radish	2,480
Sesame	4,000	Other vegetables	2,500
Clover, trifoliate	5,290	Olives	4,000
Alfalfa	5,290	Grapes	4,500
Other field crops	3,000	Figs	3,500
Tomatoes	4,590	Almonds	7,500
Squash	2,560	Peaches	7,500
Eggplant	3,510	Plums, prunes	7,500
Cucumber	2,930	Apricots	7,500
Potato	3,200	Apples	13,000
Cabbage	3,260	Pomegranates	7,500
Cauliflower	3,260	Pears	7,500
Hot pepper	3,280	Guava	7,500
Sweet pepper	2,740	Dates	13,950
Broad beans	3,180	Citrus fruit	10,860
String beans	2,310	Bananas	16,090
Peas	2,350	Other fruit	6,000
Cowpeas	2,780		
Jew's mallow	2,420		

Source: Shatanawi et al. 1998.

natural rangeland. The total quantities are increased from 1 mcm in 1994 to 1.25 mcm in 2002, as shown in Table 7-6. Fruit trees and field crops are the main consumers of irrigation water, followed by vegetables. The average annual water use amounted to 1,219 mcm over this period. To calculate the flow at the water source for other water conveyance efficiency (0.60 in Jordan's case as a lower-middle income economy country), the values in the table should be multiplied by 0.72/0.6, or a factor of 1.20, to account for the lower efficiency of 0.6. The average annual water use would be increased to 1,461 mcm. When the contribution of rangeland is added (197 mcm), the average annual water use for agricultural purposes in Jordan over the period amounted to 1,658 mcm.

TABLE 7-6. Estimated Water Quantities Used by Field Crops, Vegetables, and Fruit Trees in Jordan (mcm)

Year	Field crops	Vegetables	Fruit trees	Total
1994	340	120	589	1,049
1995	428	162	600	1,190
1996	341	103	612	1,056
1997	463	110	702	1,275
1998	486	122	717	1,325
1999	511	129	726	1,366
2000	346	120	740	1,206
2001	394	110	743	1,247
2002	391	110	753	1,254

Source: Calculations using crop-water requirement adopted in World Bank 1997 and crops produced.

Summarizing the above, we make the following conclusions:

- Average water consumption at 72% water conveyance efficiency = 1,219 mcm,
- Average water consumption at 60% water conveyance efficiency = 1,461 mcm,
- Contribution of rangeland = 197 mcm,
- Average total agricultural water consumption = 1,658 mcm,
- Average population = 4.72 million, and
- Per capita average agricultural consumption = 1,658/4.72 = 351 m³/cap.

Quantifying the Agricultural Water Trade

To quantify the volumes of all water trade flows into and out of Jordan during the period 1994 to 2002, a set of calculations has been undertaken. The basic approach is to calculate the indigenous water flow that was consumed to produce the exported commodities and the flow that would have been necessary to consume in Jordan to produce the imported quantities (agricultural shadow water). This depends on the trade figures of quantity and value as reported by the DOS (2004) and on the water duty and yield per unit area of the planted crop. Figures reported by the World Bank (1997) were used. Exported water was then analyzed to show the value added from the exports, and the share of water in that value added is assumed equal to its share in the cost of production. The value added per cubic meter of water is then calculated.

For imported water, the cost of production of the imported commodity per ton in Jordan is calculated using figures reported by the World Bank and GTZ, and the percentage of water cost is indicated. Another estimate of the water cost is made by calculating it as a percentage of the price of the imported commodity.

TABLE 7-7. Water Needed for Production of Certain Food Commodities (m³/ton)

	Hoekstra and Hung (2002)	Chapagain and Hoekstra (2003)	Zimmer and Renault (2003)	Oki et al. (2003)	Average
Wheat	1,150		1,160	2,000	1,437
Rice	2,656		1,400	3,600	2,552
Maize	450		710	1,900	1,020
Potatoes	160		105		133
Soybeans	2,300		2,750	2,500	2,517
Beef		15,977	13,500	20,700	16,726
Pork		5,906	4,600	5,900	5,469
Poultry		2,828	4,100	4,500	3,809
Eggs		4,657	2,700	3,200	3,519
Milk		865	790	560	738
Cheese		5,288			5,288

Sources: Estimates by different authors as listed in column heads.

An average of the two approaches is then taken to indicate the average cost of shadow water to Jordan.

The prices of agricultural imports and exports for the period 1994 to 2002 were obtained from the DOS and supplemented by data from other sources (Central Bank of Jordan 2004; MOA 1975). The aggregate output data set contains 28 crops: 8 field crops (wheat, barley, lentils, chickpeas, maize, onions, peas, and garlic), 13 vegetable crops (tomatoes, squash, eggplant, cucumber, potato, cabbage, cauliflower, pepper, beans, string beans, lettuce, sweet melon, watermelon), 3 citrus fruit trees (lemons, oranges, mandarins), and 4 fruit trees (grapes, peaches, apricots, and bananas). The crop-water requirements and their respective yields were quoted from the report of the World Bank (1997).

For imported commodities, the average of cereal water duty was taken as the weighted average of wheat and rice, at 1,720 m³/ton, as compared with the export average of 1,115 m³/ton. Table 7-7 presents estimates, from several different sources, of water requirements per ton of production for a variety of food commodities. This table will allow for comparison of crop requirements and adoption of the average values for noncrop commodities, such as meat, dairy products, and eggs.

Exported Water

The results of the calculation of value added per cubic meter of water, presented in Table 7-8, show that the volume of crop-related exported water averaged 68.61 mcm/y, representing about 93% of all exported water over the period 1994 to 2002 for the 28 selected crops. The total average exported water is therefore 74 mcm. The water consumed in the production of the exported field crops represents only 2.61%, whereas for other fruits the value is about

TABLE 7-8. Exported Water for Selected Crops, 1994–2002

	Field crops	Vegetables	Citrus	Fruit trees	Total/ average
Total export (tons)	1,606	315,062	37,454	2,294	356,416
Value of export (US$ million)	0.8	86.1	11.9	2.1	100.9
Price of export (US$/ton)	486.8	273.4	317.3	895.6	283.0
Cost of production (US$/ton)	150	197	91.3	166.8	185.5
Percent cost of water	52	6.08	24.13	28	7.3
Value added (US$/ton)	336.8	76.4	226	728.8	97.5
Share of water in value added (US$/ton)	175.1	4.645	54.53	204.06	7.12
Water used in exported commodities[a] (m^3/ton)	1,115	170	313	670	192.5
Value added of water (US$/m^3)	0.30	0.45	0.72	1.08	0.50
Exported water (mcm/y)	1.790	53.56	11.72	1.537	68.61
Cost of exported water (US$ million)	0.089	2.678	0.586	0.079	3.432
Cost of water per ton of produce (US$)	55.41	8.5	15.65	34.44	9.63

a. Numbers extracted from World Bank 1997, vol. 2: 1–6.

2.24%. Water in exported citrus products represents about 17.08%. The main category of agricultural products exported is vegetables, and the water consumed in their production represents 78.07% of exported water. Measured in per capita share, the exported water amounted to 15 m^3/y. The revenue for irrigation water in the Jordan Valley is about US$0.021/m^3 (15 fils; collection rate 10.4 fils), compared with the real cost of US$0.050/m^3 (35 fils). The table shows the data related to the selected 28 crops, which on average represented 93% of the total exported agricultural commodities over the investigation period. The cost of water used to produce them is based on the real cost, or US$0.050/m^3.

The value added in the export of goods also is shown in the Table 7-8. It indicates that the least attractive value added per cubic meter of water is the case of field crop exports, which averaged US$0.30/m^3. The most attractive export as far as water is concerned is the fruits, where the value added per cubic meter of water averaged US$1.08, followed by citrus at $0.72 and vegetables at an average of $0.45. In all cases, however, the value added per cubic meter of water, taken as a weighted average of the production cost percentages, is many times the actual cost of irrigation water used to produce the exported commodities.

Imported Water

The same procedure described above is applied to estimate the volume of water flows into Jordan brought about by the imported crops. The results, presented

TABLE 7-9. Agricultural Shadow Water and Costs for Selected Crops, 1994–2002

	Field crops	Vegetables	Citrus	Fruit trees	Total
1. Total imported quantities (tons)	1,464,428	19,145	22,829	9,301	1,515,703
2. Values of imports (US$ million)	225.6	6.9	10.6	7.2	250.3
3. Price of imports (US$/ton) = 2 / 1	154.0	360.0	464.4	774.1	165
4. Indigenous water saved[a] (m^3/ton)	1,720	170	313	670	1,685
5. Imported water (mcm/y) = 4 × 1	2,519	2.5	18.7	14.8	2,555[c]
6. Cost of production[b] ($/ton)	150	197	91.42	166.8	185.5
7. % cost of water[b]	52	6.08	24.13	28	7.3
8. Cost[a] (US$ million) = 7 × 6 × 1	114	0.23	0.50	0.43	20.52
9. Water as % of revenue	16	4.3	6.95	5.2	4.8
10. Cost[b] = 9 × 2 (US$ million)	36.1	0.3	0.73	0.37	37.5
11. Average water cost (US$ million) 8 + 10 / 2	75	0.26	1.23	0.80	29.0
12. Cost per cubic meter (US$) 11 / 5	0.0297	0.104	0.065	0.054	0.0113

a. Field crops figure is weighted average of wheat and rice, ratio 3:1; other figures calculated from World Bank 1997, Annex C.
b. Calculated from tables in World Bank 1997, 2–3, attachment 2.
c. Adjusted to 2,668 mcm by adding the shadow water of the rest of the imports at 113 mcm.

in Table 7-9, show that the crop-related imports saved Jordan a total of 2,555 mcm as an annual average over the period of analysis, 1994 to 2002. An amount of 113 mcm is added to it to account for the remainder of crops not included in the analysis, and the total average becomes 2,668 mcm. The imported field crops accounted for 97% of total imported water.

The selected 28 crops whose data is shown in Table 7-9 represent, on average, 90% of the total imported agricultural commodities over the investigation period. The total tonnage becomes about 1,684,114. Assuming the remaining 10%, or 168,411 tons, are of fruit products, with about 670 m^3/ton of water used to produce them (Table 7-8), the additional imported water would amount to 113 mcm, making the total indigenous water saving from crop imports equal to 2,555 + 113, or 2,668 mcm/y of shadow water. The net crop shadow water would be this total shadow water minus the 74 mcm of export water, or 2,594 mcm/y.

The Cost of Agricultural Shadow Water

Two approaches were followed to determine the cost to Jordan of agricultural shadow water, which is the water quantity that would otherwise have to be used to produce the imported agricultural commodities.

a. Water cost as a percentage of the cost of production of the commodity in Jordan was taken as a base. The cost of production, as taken from Annex C of the World Bank's 1997 report, is shown in row (6) of Table 7-9, and the water percentage of that cost is shown in row (7).

b. Water cost as percentage of revenue in Jordan was taken as a base. The percentage of revenue is based on data of Annex C (World Bank 1997), and is shown, in Table 7-9, in row (9) whereas the revenue is shown in row (2).

The cost of water for case (a) is shown in row (8) and for case (b) in row (10). The average water cost is shown in row (11). The average cost of water amounted to US$29 million, and its average quantity was 2,555 mcm (5). The average cost of imported water is therefore US$0.01135/m^3, or 8.05 fils/m^3. Compared with the actual cost of irrigation water in the Jordan Valley of US$0.05, or 35 fils, the cost of imported water is many times cheaper, about 23% of the actual running cost of irrigation water.

Other Food Commodities

In addition to plant crops, Jordan trades other food commodities. According to the DOS database, the main ones imported into Jordan are predominantly live animals, meats and meat products, cereals and maize, oil crops, vegetable oils, sugar, pulses (legumes grown for dry grain), and starchy roots. Each category includes a variety of crops and related products.

Water demand to produce a unit weight of the imports varies among the commodities. Water demand indices for live animals are shown in Table 7-10. Indices for the other commodities listed above were obtained mainly from Hoekstra (2003) and Chen et al. (1995). Data for volumes and values of the commodities imported are taken from DOS (2004). The detail of the accounting is specified below.

For the purpose of calculating the shadow water of products, Zimmer and Renault (2003) distinguish among primary products (crops), processed products (such as sugar, vegetable oil, and alcoholic beverages), transformed products (including animal products), by-products (such as cottonseeds), multiple products (such as coconut trees), and low- or nonwater consumptive products. Table 7-11 shows the products used in the analysis to add to the crop products of Table 7-9. The average estimates of water indices given in Hoekstra 2003 and Hoekstra and Hung 2002 are used to estimate shadow water of the selected food items. The average annual volume of shadow water of these non-

TABLE 7-10. Water Demand Indices for Live Animals (m³/ton)

	Horses	Sheep	Goats	Bovine	Swine	Dairy cattle	Egg layers	Poultry
Australia	11,707	6,343	6,585	11,707	6,117	1,213	4,053	2,373
Canada	9,619	5,666	5,440	9,619	3,268	823	2,314	1,358
China	11,186	5,940	10,016	11,186	2,160	2,079	8,651	3,111
India	12,729	6,589	11,237	12,729	4,175	2,596	23,692	8,499
Ireland	7,575	5,246	4,809	7,575	2,012	715	1,544	908
Italy	9,581	5,710	5,407	9,581	3,459	842	2,792	1,637
Japan	10,751	5,786	6,105	10,751	4,325	1,113	3,488	2,044
Korea, PDR	11,116	5,926	8,096	11,116	3,526	1,597	6,874	2,860
South Korea	13,172	6,735	9,572	13,172	6,685	2,171	13,668	5,679
Netherlands	7,680	5,261	4,823	7,680	2,086	730	1,555	914
Russia	12,310	6,495	9,055	12,310	5,488	1,967	11,312	4,702
United States	10,056	5,715	5,592	10,056	3,371	827	2,222	1,304
Average	10,624	5,951	7,228	10,624	3,889	1,389	6,847	2,949

Sources: Compiled from Hoekstra (2003) and Hoekstra and Hung (2002).

TABLE 7-11. Distribution of Values, Quantities, and Shadow Water among Selected Food Groups in Jordan, Average for 1994–2002

External trade	Import: value (US$ million)	Import: quantity (tons)	Export: value (US$ million)	Export: quantity (tons)	Shadow water (mcm)	Water exported (mcm)	Net shadow water (mcm)
Live animals	39.9	34,175	38.9	21,467	247	127	120
Meat and meat products	98.0	31,199	1.0	521	456	3	453
Sugar	58.8	185,269	0.1	181	357	0	357
Dairy and poultry	66.1	28,501	8.2	32,606	153	92	61
Coffee and tea	30.8	15,351	0.9	1,049	28	2	26
Fats and vegetable oils	91.9	166,847	3.3	3,399	978	23	955
Total[a]	385.5	461,342	52.4	59,223	2,219	247	1,972

a. Represents only 71% of the volume annual trade average in these commodities.
Sources: Hoekstra 2003 and Hoekstra and Hung 2002.

crop commodities should be adjusted upward to account for the entire average volume of trade. The adjusted shadow water becomes 3,125 mcm, the exported water would be 348 mcm, and the average net shadow water becomes 2,777 mcm.

Total Agricultural Shadow Water

The total net shadow water volume in Jordan as an average for the years 1994–2002 would be the sum of the crop and noncrop net import effect:

$$W_{SH} = \text{Crop water} + \text{noncrop water.}$$

The two components forming the shadow water were calculated above and shown to be 2,594 and 2,777 mcm, respectively. Their total is expressed by:

$$W_{SH} = 2,594 + 2,777$$
$$= 5,371 \text{ mcm,}$$

where

W_{SH} is the total of the agricultural shadow water, crop and noncrop.

The shadow water is calculated through another approach, using the crop-water requirements as adopted in the World Bank's 1997 report for crops, displayed in Table 7-6, and the average water indices of Table 7-7 for items not included in that report (noncrop items), applied to the respective crop areas planted between the years 1994 and 2002 and to the noncrops of the same time span. The flow of shadow water throughout the study period is shown in Table 7-12.

The net shadow water of this second approach, as shown in Table 7-12, is 5,445 mcm, compared with a total of 5,371 calculated in the first approach above. The average population during the nine years of analysis is 4,724,245 persons. Hence, the per capita shadow water is 1,133 m^3/cap/y using the first approach or 1,148 m^3/cap/y using the second, with an average of 1,140 m^3/cap/y.

A third estimate of shadow water can be obtained from the theoretical calculation by which the total average per capita need for agriculture in a lower-middle income economy such as Jordan was shown to be 1,500 m^3. The net indigenous water used in agriculture should be subtracted from this average to get the net agricultural shadow water. The agricultural water used as an average in the period of analysis is shown in Table 7-6. The average of water used during the nine years is 1,219 mcm/y, yielding an average per capita use of 258 m^3/y. Exported water average was 15 m^3/cap/y (see above in the Exported Water section). The net indigenous water use was therefore 243 m^3/cap/y.

TABLE 7-12. Estimate of Agricultural and per Capita Water Flow, 1994–2002

Year	Shadow water of imports[a] (mcm)	Exported water (mcm)	Shadow water balance (mcm)	Cost of imports (US$ million)	Population	Per capita shadow water (m³)	Per capita exports water (m³)	Per capita net balance of water (m³)
1994	6,234	595	5,639	845.9	4,139,458	1,506	144	1,362
1995	5,297	381	4,917	737.5	4,291,000	1,235	89	1,146
1996	6,652	623	6,029	904.3	4,444,000	1,497	140	1,357
1997	6,267	864	5,402	810.4	4,600,000	1,362	188	1,174
1998	5,897	671	5,225	783.8	4,755,750	1,240	141	1,099
1999	5,883	625	5,258	788.7	4,900,000	1,201	128	1,073
2000	5,925	446	5,479	821.8	5,039,000	1,176	89	1,087
2001	5,952	371	5,581	837.1	5,182,000	1,149	72	1,077
2002	5,777	302	5,476	821.3	5,329,000	1,084	57	1,028
Average	5,987	542	5,445	817.0	4,742,245	1,262	114	1,148

a. To be adjusted upward by 20% to account for the difference in the estimated irrigation efficiencies when comparison is to be made with the theoretical formula.

Source: Compiled from Table 7-11 and the DOS data on export and import of agricultural commodities for the years shown.

Accordingly, net shadow water can be estimated at 1,500 − 243 = 1,257 m³/cap/y. This is to be compared with the shadow water shown in Table 7-4 as 1,248 m³/cap for the year 2002, which includes the export water use. When adjusted for export, the shadow water becomes 1,263 m³/cap.

The difference between the nine-year averages calculated theoretically (1,257) and the actual conditions of shadow water (1,240) is 17 m³/cap, or 1.37%.

That the factual conditions and the theoretical calculation of shadow water differ can be explained in one of two ways:

a. The theoretical estimate was based on a certain value of caloric intake per day, split between animal- and plant-based protein. The deviation from it could be attributed to a similar deviation in the daily caloric intake per capita.
b. The efficiency of water use in the theoretical analysis is 0.6 for a lower-middle income country. Any deviation from it entails deviation from the calculated water threshold and, consequently, the shadow water.

Let's look at another year for comparison. In 2000, the theoretical calculation shows that agricultural shadow water in Jordan was 1,180 m³/cap. However, the actual data in Table 7-12 show that the shadow water was 1,087 m³/cap, which is a difference of +93 m³/cap/y, a margin of error of 8%. Again, a slight increase in the assumed irrigation efficiency in the theoretical calculation can eliminate the difference. It can thus be said that the theoretical calculation of agricultural shadow water corresponds reasonably well with the actual shadow water observed in the real market.

Summary

A method was introduced to measure the blue-water equivalent of green water and was found to average 866 mcm in Jordan. This is more than the surface-water resources available to the country and warrants a closer look at management and yields through a focus on blue water. The country's natural renewable freshwater resources, including green water, average 1,970 mcm (Chapter 1), compared with its equilibrium needs of 9,100 mcm for a population of 5.35 million people (at 1,700 m³/cap/y, as per Table 7-4), constituting a 21.6% sufficiency ratio. Augmentation with gray and silver water slightly improves the sufficiency ratio.

The imbalance in the population–water resources equation is a major handicap for Jordan. Poverty in energy resources makes the water situation even more critical. An expenditure of energy is necessary for water use, and Jordan imports practically all of its energy needs from the outside. Energy costs impact the cost of water service and directly affect the water tariff. The gaps in the water equation and energy supply are bridged by commodity imports. The analysis of shadow-water cost indicates that, for Jordan, it is much cheaper to import agricultural commodities than to produce them using the scarce water resources. This brings up the question of whether it would be more prudent to divert fresh agricultural water to urban use and replace the agricultural production with imports. A host of social and environmental issues must be addressed before such a question can be answered.

The framework presented in this chapter suggests that policymakers evaluate the relative scarcity of key resources in the context of setting national goals for economic and social development, achieving food security, and improving the quality of life.

References

Allen R., L. Pereira, D. Raes, and M. Smith. 1998. *Crop Evapo-transpiration: Guidelines for Computing Crop Water Requirements.* FAO Irrigation and Drainage Paper 56. Rome, Italy: Food and Agriculture Organization.

Central Bank of Jordan. 2004. *Monthly Statistical Bulletin* 40: 66.

Chapagain, A.K., and A.Y. Hoekstra. 2003. Virtual Water Trade: A Quantification of Virtual Water Flows between Nations in Relation to International Trade of Livestock and Livestock Products. In Hoekstra 2003, 57.

Chen Y., G. Guo, G. Wang, S. Gang, H. Luo, and D. Zhang. 1995. Water Demand and Irrigation of Main Crops in China. Beijing, China: Hydropower Publishing House.

Critchley, W., and K. Siegert. 1991. *Water Harvesting: A Manual for the Design and Construction of Water Harvesting Schemes for Plant Production.* Rome, Italy: Food and Agriculture Organization.

Doorenbos, J., and W. Pruitt. 1992. *Crop Water Requirements.* FAO Drainage and Irrigation Paper 24. Rome, Italy: Food and Agriculture Organization.

DOS (Department of Statistics). 2004. *Statistical Yearbook.* Amman, Jordan: DOS.

Elmusa, S. 1997. *Water Conflict: Economics, Politics, Law and Palestine-Israel Water Resources.* Washington, DC: Institute for Palestine Studies.

FAO (Food and Agriculture Organization). 1999. *Production Year Book.* Cited by Haddadin 2003a.

———. 2000. Food Balance Sheets. http://apps.fao.org/page/collections?subset= nutrition (accessed October 6, 2005). Data was not available for Iraq, Qatar, Bahrain, and Oman.

———. 2004. Yields of Crops in World Countries. http://apps.fao.org/page/form?collection=Production.Crops.Primary&Domain=Production&servlet=1&language=EN&hostname=apps.fao.org&version=default (accessed October 6, 2005).

Gleick, P. 2000. *The World's Water: The Biennial Report on Freshwater Resources.* Washington, DC: Island Press.

———. 2004. Personal communication with Dr. Munther Haddadin, October 13, at the Vatican.

Gleick P., W. Burns, and E. Chaleckiet. 2002. *The World's Water: The Biennial Report on*

Freshwater Resources, 2002–2003. Pacific Institute for Studies in Development, Environment and Security, Oakland, CA. Washington, DC: Island Press. Source data extracted from Table 2, Freshwater Withdrawals by Country and Sector, 2002 update, 245–51.

Haddadin, M. 2002. Water Issues in the Middle East: Challenges and Opportunities. *Water Policy* 4(3): 205–22.

———. 2003a. Significance of Shadow Water to Irrigation Management. Paper sponsored by the University of Oklahoma, Norman, and Oregon State University, Corvallis, submitted to Prince Sultan Bin Abdulaziz International Prize for Water.

———. 2003b. Exogenous Water: A Conduit to Globalization of Water Resources. In Proceedings of the International Expert Meeting on Virtual Water Trade. Research Report Series no. 12, 159–69. Delft, The Netherlands: IHE.

Hjort, K., M. Zakaria, and F. Salah. 1998. *An Introduction to Jordan's Agriculture Sector and Agricultural Policies.* Access to Microfinance & Improved Implementation of Policy Reform (AMIR Program). Amman, Jordan.

Hoekstra, A. 2003. Virtual Water Trade. In Proceedings of the International Expert Meeting on Virtual Water Trade. Value of Water Research Report Series no. 12, 12. Delft, The Netherlands: IHE.

Hoekstra, A., and P. Hung. 2002. Virtual Water Trade: A Quantification of Virtual Water Flows between Nations in Relation to International Crop Trade. Value of Water Research Report Series no. 11, Appendix I, 1–12. Delft, The Netherlands: IHE.

Johnson, G. 2003. Deep in Universe's Software Lurk Beautiful, Mysterious Numbers. *New York Times,* May 20, D3.

Merett, S. 2003. Virtual Water—A Discussion: Virtual Water and Occam's Razor. *Water International* 28(1): 103–15.

MOA (Ministry of Agriculture). 1975. *Experiments Conducted by the Department of Research and Extension.* Amman, Jordan: MOA.

MWI (Ministry of Water and Irrigation). 2004. *Water Resources in Brief.* Amman, Jordan: MWI.

Oki, T., M. Sato, A. Kawamura, M. Miyake, S. Kanae, and K. Musiake. 2003. Virtual Water Trade to Japan and in the World. In Hoekstra 2003.

Postel, S. 1999. *Pillars of Sand.* New York: W.W. Norton & Company.

Shatanawi, M., G. Nakshabandi, A. Ferdous, M. Shaeban, and M. Rahbeh. 1998. *Crop Water Requirement Models for Crops Grown in Jordan.* Technical Report no. 21. Amman, Jordan: University of Jordan, Water and Environmental Research and Study Center.

World Bank. 1997. *The Hashemite Kingdom of Jordan: Water Sector Review.* World Bank Report no. 17095-JO. October 15. Washington, DC: Rural Development, Water and Environment Department, Middle East and North Africa Region.

Zimmer, D., and D. Renault. 2003. Virtual Water in Food Production and Global Trade: Review of Methodological Issues and Preliminary Results. In Hoekstra 2003.

Notes

1. The income categories are those specified by the World Bank. The low income category has a per capita share of the gross national income of $755 per annum or less, lower-middle ranges between $756 and $2,995, upper-middle ranges between $2,996 and $9,265, and high income is $9,266 or more.

2. One dunum is one-tenth of a hectare, or 1,000 square meters.

3. Credit goes to Professor Tony Allan of the School of Oriental and African Studies at the University of London for introducing the term *virtual water* in reference to the water "embedded" in the imported food commodities. Dr. Munther Haddadin had an elaborate discussion with Professor Allan in November 1991 of the concept he introduced and kept communicating with him thereafter.

4. *Merriam Webster's Collegiate Dictionary,* 7th ed.

5. The terms *strain* and *stress* are borrowed from the science of engineering mechanics. In its simplest form, *strain* is defined as the ratio of the elongation of contraction of a rod, subject to axial force, to its original length before that force was applied. In the simplest form illustrated above, *stress* is defined as the force per unit area of the rod's cross section.

6. An unspecified portion of this water was used for cooling, and thus not all use was consumptive.

7. In some cases, electricity is generated in Canada and transmitted to the United States. Examples include Niagara Falls and hydropower generation in British Columbia with transmission as far south as California.

8

Water and Wastewater Management

Governance and Policy Framework

Munther J. Haddadin, Ra'ed Daoud, and Helena Naber

Jordan has been undergoing water strain as a result of water stress (number of persons per unit flow). In 2004, for example, 2,717 persons competed for a unit flow of blue and green water, compared with the equilibrium condition of 588 persons per unit flow (equilibrium need is 1,700 m^3/cap/y). This means that water stress in Jordan, in 2004, was 4.6 times that at equilibrium. And the water stress will be rising over time, resulting in more strain than has been experienced thus far.

Such a tight water situation creates many challenges for water managers and the government at large, such as determining the water allocation criteria and implementing policy. Primary issues of governance include stakeholder participation, private-public sector partnership, involvement of nongovernment organizations (NGOs), and the format of government administration, among other issues of significance and importance.

At the heart of these challenges is how to optimize the national economic, social, and environmental benefits from the scarce water resources. Unlike water-rich countries, whose water stress at equilibrium amounts to 833 persons per unit flow, water-strained countries cannot afford trial-and-error approaches, nor can they afford mistakes in water planning and management. Their water officials have to be on the alert most of the time, looking for ways and means to exercise prudent management, minimize adverse impacts, and maximize gains. Water managers therefore have to be trained and qualified to be able to make the right decisions at moments of crisis, and make them quickly in many circumstances. The situation is exacerbated when water stress is coupled with financial stress, as has been the case in Jordan for the past three decades. Worse yet, as also is the case in Jordan, is when the country is almost void of energy resources and is experiencing both water and financial stress.[1]

In this chapter, we examine the issues of governance of water and wastewater, private sector participation in utilities management, government subsidies in the water sector, stakeholder participation, and users' associations. The policy trend in letting out management contracts to private sector companies is explained and the ongoing contracts appraised. Forthcoming management contracts or concessions are also addressed.

Legal and Institutional Aspects of Water Governance

Perhaps the most difficult of the tough water challenges facing Jordan is the challenge of water governance, an integrated set of issues to be addressed in successful water management. The legal and institutional aspects of water resource management and administration were reviewed in Chapter 2. Until municipal-water hardships started to surface in the mid-1960s, water management had been the responsibility of elected municipal councils in Jordan. As such, these councils represented the stakeholders of water and wastewater services.

As the imbalance in the water resources–population equation increased, the central government, through the Natural Resources Authority (NRA), took over the responsibilities of the municipal councils in 1966. Only the capital city of Amman was outside this arrangement, and its council, appointed at the time, continued to exercise its role in the provision of water and wastewater services. At that time, the NRA thought of involving the beneficiaries of municipal water in water management, in a move that represented the first attempt to implement sound water governance. A water council was formed for the governorate of Jerusalem in 1966, for which a major municipal water project, financed by the International Development Association (IDA) of the World Bank Group, was under way. The NRA was not empowered by its law to form water councils, nor did that law refer to any role for the beneficiaries. The move was credited to the dialogue between an enlightened NRA official and the field missions of IDA. Water councils were to be formed under various legislations, especially the law that created the Water Authority of Jordan (WAJ). However, the formation of these councils was not honored.

Current Water Administration

As outlined in Chapter 2, administration of water resources and distribution to beneficiaries took the form of institutional centralization at the beginning, shifted to pluralism and decentralization, then was centralized again. The central authority for water resources, since 1988, has been vested in the Ministry of Water and Irrigation (MWI), with its constituent authorities, the WAJ and the Jordan Valley Authority (JVA). A third branch of the MWI, called, confusingly, the Ministry, was created in 1992 to be responsible for studies and planning and to follow up with donor agencies

As an autonomous statutory body, the WAJ is responsible for water and wastewater services in the country. It has its own independent treasury; its own procurement, financial, and personnel regulations to permit flexibility; and its accounts organized according to recognized accounting methods and audited by qualified auditors from the private sector. The JVA is responsible for irrigation-water uses, including dams. The JVA and the third branch are part of the government administration system, and their accounts are audited by the government's Audit Bureau. Their budgets are allocated by the government, and any revenue they generate goes straight to the state Treasury.

The minister of water and irrigation chairs both the Board of the WAJ and the Board of the JVA and runs the third branch. Coordination among the three entities is assured through the minister and the two boards, about half of whose members are the same officials. Representation of stakeholders is guaranteed in the two boards.

Under Article 28 of the WAJ law, the Council of Ministers, upon the recommendation of the minister for water and irrigation, may assign any of the authority's duties or projects to any other body from the public or private sector or to a company owned totally by the WAJ. This may include transfers of project management, a lease, or ownership of projects. When the capital projects in municipal water and wastewater multiplied in the latter part of the 1990s, and in response to a desire by the donors, a Project Management Unit (PMU) was created within the WAJ to monitor the various capital assistance projects, including the management contracts for water and wastewater. The PMU exists as an organizational unit with its own board of directors, which reports to the minister for water and irrigation, the chairman of the board of the WAJ. It was envisaged that the PMU would take on responsibility for monitoring private sector participation contracts, such as the As-Samra build-operate-transfer (BOT) contract and the northern governorates' management contract.

Arrangements such as the creation of the PMU have been made as a result of consultation with the donors to municipal water projects. Despite the autonomy given by law to the WAJ, the sheer mass of the organization imparts static inertia, impeding its flexibility. Amendment of regulations by the government introduced constraints on free administrative and financial action, and the frequent turnover of ministers did not help maintain the momentum of efficiency. Donors insisted on efficiency, less bureaucracy, and flexibility of action to ensure speedy implementation of the financed projects. The government abides, and when new organizational arrangements, such as PMU, are made, they typically experience a "honeymoon" phase that lasts for a few years before bureaucracy moves in. Efficiency and excellence usually attract conservative reactions of the type that prompts government to gradually curtail the measures of flexibility granted to the new organization. This has been the case with the various authorities created in Jordan since the first was established in 1959.

The creation of the third branch of the MWI was prompted by the initiation of the Middle East peace process. The minister realized that a full-time team, kept away from everyday operational chores, was needed to come up with visions for water resource projects in the context of regional cooperation and perform initial studies before such projects were seriously considered. The third branch was created for this purpose and operated under the minister and a new secretary general. The division of labor among the three branches was to be done in accordance with the laws of both the JVA and the WAJ, which require authorization of each for any third party to do any task within their responsibilities. In actual implementation, it became evident that the third branch performed the tasks assigned to it without due regard to authorization, taking cover under the minister against any criticism concerning legalities. An urgent need currently exists to clearly define the responsibilities of the WAJ and the third branch to conform to legal stipulations in their respective legislations.

A handicap of the current arrangements and other water administration issues is the rate of change of water ministers. Between the ministry's establishment in 1988 and the year 2006, the position of minister has changed hands 17 times, an average period of service of one year per minister. This turnover undoubtedly dilutes the effectiveness of the minister and puts at risk the continuity of policies. Fortunately, the position of secretary general of each of the three branches of the MWI witnessed less frequent turnover: five times for the JVA, three for the WAJ, and five for the third branch, the Ministry.

Irrigation Water and Land Redistribution Policies

In the Jordan Valley, prior to government involvement in irrigation infrastructure, private farmers provided their own means of irrigation. Farmers used the waters of the side wadis to irrigate their lands in accordance with the water rights registered in their title deeds. Outside the Jordan Valley, development of irrigation, whether through diversion of perennial streams or abstraction from groundwater, has been the responsibility of the private farmers.

Government's involvement in irrigation, as a formal policy, was prompted by the need to utilize water resources, create jobs, implement economic and social development, and protect Jordan's share of water resources in the Yarmouk River. Capital assistance was granted by the U.S. government in 1958 to build the East Ghor Canal and its distribution networks (Chapter 4). Government policy, outlined in cooperation with the donor, has been to take on the responsibility for irrigation infrastructure up to the farm gates, after which the farm owners undertake to build on-farm water distribution systems. Thus the construction of dams, diversion structures, conveyance and distribution systems, and all the appurtenant works has been the responsibility of government. Hence society, represented by the executive branch, in effect became a partner in the development process, especially as the recovery of the government investment was not on the table. This provided the moral justification to provide for a land redistribution package in the legislation, by which owners of land in the project would retain part of their property and the remainder would be redistributed to landless farmers.

On-farm agriculture has always been the responsibility of the landowner, who could rent it out to tenants or enter into sharecropping agreements with them. This form of partnership in the Jordan Valley proved to be a success from the social, economic, and environmental points of view. The Agricultural Credit Corporation, an autonomous organization of the central government, has been active in providing capital and seasonal loans to farmer–owners to assist them in developing their farms.

The government's policy on stakeholder participation was stipulated in the laws that organized its own irrigation undertakings. Two members of the Board of Directors were appointed by the Council of Ministers to represent the stakeholders. They participated with the other members, who were representatives of other government departments, in overseeing the plans, approving the budgets, and following up on the projects. The Farmers' Selection Committee, which selected farmers to receive redistributed land and helped them settle into the new farm units, consisted of three members, one of whom was an appointed representative of farmers in the area.

Issues of Governance

In addition to the legal and institutional aspects, several important issues enhance water governance in Jordan today. These include participation by stakeholders and the private sector and enabling the poor to afford and obtain water and wastewater services.

Stakeholder and Private Sector Participation

Each authority institution includes in its Board of Directors two members appointed by the Council of Ministers to represent the stakeholders. The extent of stakeholder participation has been limited to this membership; although participation was meant to be expanded, as the Water Authority Law stipulates, no material steps were taken to translate legislation into action. Stakeholder participation is likely the result of importing ideas from abroad. Sharing of authority with government has not been one of the virtues of Arab culture.

Consultation, however, is enshrined in that culture, but decisions are made by the ruler.

More recently, prompted by the economic restructuring program, the government has embraced some new techniques for managing the water and wastewater services. A contract has been in effect since 1999 for the management by a private consortium of water and wastewater services in the Greater Amman area.

Stakeholder participation in the management of irrigation water is being attempted in the Jordan Valley, where the farmers of a given irrigation lateral are assigned the responsibility for its operation.[2] The idea was introduced by the World Bank and is slated for expansion through the establishment of the Water Users Association.

Access for the Poor

Issues of enabling the poor to afford water and wastewater services were addressed in the water tariff system. A block tariff has always been the means to enable the poor, with their modest consumption subject to subsidy. In the most recent tariff structure, if the actual consumption by low income people exceeds the estimated consumption of 20 m^3/subscriber/quarter, the water subsidy is lifted off the entire consumed quantity. Consumers of higher amounts never benefit from the subsidy earmarked for the poor who consume less than the above prescribed amount per quarter.

In irrigated agriculture, subsidized loans are advanced to landowners with priority given to small farmers. Subsidies were built into the water tariff that benefit all farmers across the board. Issues of equity were addressed by the top water administration at the level of the secretary general and the minister. Under the water strains the country has been experiencing, a rotation system of service has been prepared for every summer season and announced to the public through the written media before its application. Poor neighborhoods receive their equitable turn in the rotation. In fact in Amman, the poorer neighborhoods happen to be served better-quality groundwater, whereas the more affluent neighborhoods of western Amman receive mostly surface water treated with chemicals.

Water networks now reach more than 97% of the population of the country, and the poor neighborhoods of all cities are connected. The same cannot be said about wastewater networks, however, although their coverage is better in the poor neighborhoods than in the wealthy.

Challenges Facing the Water Sector

The water sector in Jordan today still faces several challenges, and addressing them will be no easy task for the government. Some of these challenges include the need to provide reliable water supplies, decrease the unsustainable use of groundwater, and protect water resources from pollution. Other issues are allocation of water resources, water governance, institutional framework, increasing costs of energy, and the financial capacity and cost recovery of water and wastewater utilities.

The Need to Secure Reliable Water Supplies

Jordan's surface-water and groundwater resources are almost fully developed to supply various users, and little freshwater potential remains for development. The increased use of treated wastewater will enable the diversion of more irrigation freshwater resources to municipal use, and thus this has a high priority. Desalination of brackish water has been adopted to supply Amman with potable water, and more desalinated water is likely to be used in Jordan in the future.

Unsustainable Use of Groundwater

The unsustainable use of groundwater resources in Jordan represents one of the major challenges. Jordan's renewable groundwater resources are being abstracted by more than double their safe yield capacity. The costs associated with this practice far outweigh the accruing benefits. The eventual cost will be losing the aquifers that are subject to overabstraction, because of water quality degradation or because the sources dry up, or both. The success of the WAJ in enforcing the provisions of the groundwater control legislation has been modest thus far.

Environmental Concerns

Pollution of water resources is a standing concern. Surface-water streams are exposed to pollution from point-source and nonpoint-source pollutants. Some pollution originates in the neighboring countries, primarily in the Yarmouk catchment inside Syria. Agricultural inputs into the catchment are sources of pollution, and untreated wastewater is frequently another source. From Israel, treated and untreated wastewater is a source of pollution to the Jordan River south of Lake Tiberias.

The overabstraction of groundwater from an aquifer is an invitation to groundwater from an adjacent aquifer, possibly brackish, to intrude into that aquifer, causing quality degradation. With the increasing reliance on desalinated water, disposal of the brine resulting from the process is an emerging environmental burden.

Allocation of Resources

Water allocation priorities are recognized in Jordan's water strategy and related policies. In the allocation of new water resources, first priority is given to municipal and industrial uses, and a lower priority is given to agriculture. No attention is paid in the strategy to environmental water, a use that is gaining increasing importance. The abstraction of groundwater resources should be curbed such that the total abstraction from each aquifer matches the average recharge. Greater use of gray-water resources will help in the process of reallocation. Depending on the circumstances, some water allocation to the environment may be possible.

Water Governance

Good water governance requires effective sociopolitical and administrative measures that are based on an integrated approach to water resource management with clear accountability procedures. Technical, institutional, managerial, legal, and operational activities should be simultaneously addressed, and transparent and participatory processes should account for beneficiaries' interests and ecological and human needs.

Jordan has covered considerable ground in developing a clear vision for water resource management at national and provincial levels. What remains to be achieved is an integration of the vision in a management system that develops lateral and vertical linkages with the sectors most affected by water policy. More urgent are the needs for transparency and the participation of stakeholders, including NGOs, in decisionmaking, especially when it comes to decisions on water and wastewater resource development. In this regard, Jordan has much yet to be adopted and implemented.

Institutional Framework

Improvements to the public water institutions undertaken thus far have addressed Treasury subsidies, systems efficiency, and delivery of services. Various options have been explored concerning public-private partnerships.

The institutional arrangements in the water sector have vacillated between segregation and integration and between local and central governments, as described in Chapter 2. The country now has a store of experience and heritage in systems of water administration. A role for the private sector has been carved in the form of management contracts, but water privatization options have not been entertained, nor will they in the near future. Water is public property as stipulated by law, and privatizing it would jeopardize its ownership by the public. All Muslim societies observe the public ownership of water as stipulated by the hadith of Prophet Mohammad, "People are partners in three: water, pastures, and fire."

On the other hand, the role that water plays in the economy demands greater efficiency in the provision of service. This is more the case when the country is facing economic challenges and undergoing an adjustment program. Maximum service should be attained per dinar spent in its provision. Services provided by the government are subject to rules and regulations that control public spending, but these are hardly fit to cope with the flexibility and promptness needed to operate water and wastewater systems. Additionally, the requirements that the government budget deficit be reduced to manageable limits dictate that subsidies on consumer goods—and water is one—be gradually lifted and eventually eliminated.

Energy

The provision of water and wastewater services is invariably linked to the consumption of energy in its various forms. Electrical energy is consumed to pump water into the water networks, and potential (gravity) energy drives the flow of water and wastewater in their courses. Wastewater treatment uses energy in solar or mechanical form, depending on the treatment technology. The cost of the provision of service is thus sensitive to the cost of energy. Jordan imports all its needs of fossil fuel, refines it to different grades, and uses fuel oil to generate most of its electricity.

The efficiency of the provision of water and wastewater services therefore depends on both indigenous and exogenous factors. Energy accounts for more than a third of the cost of the provision of water services in Jordan. When the indigenous factors account for the rest of the cost, it becomes doubly challenging to boost the efficiency of services to yield high returns for every dinar spent. Escalation of market prices of fossil oil heavily impacts the cost of provision of water services.

Financial Capacity and Cost Recovery

In countries sustaining high water strain, the marginal cost of water is usually high, because the nearby water resources, which are less expensive to develop, have already been put to use, and additional water has to be brought from locations farther away at higher development costs.

A healthy economy frees the government Treasury from having to provide consumer subsidies. Jordan has been implementing a process of gradually lifting subsidies, although its GDP places the country in the lower-middle income category. The Treasury income currently meets the recurrent government expenditure, with a modest surplus for capital investment. Loans are necessary to bridge the wide gap between capital investment needs and available indigenous revenues.

Jordan's GDP hardly enables its population to afford meeting the full cost of water and wastewater services. Cross-subsidies have been embedded in water tariffs, but the revenue falls short of full cost recovery. Chapter 6 shows the rates of full cost recovery of municipal water and wastewater services and of irrigation services.

It has been recommended that the water and wastewater cost should not exceed 2% of the GDP. The challenge Jordan is facing is formidable: to boost the GDP so that 2% of it would meet the requirement of full recovery of water and wastewater services. When the exogenous cost factors are considered, especially components of operation and maintenance, the challenge intensifies. The cost of provision of municipal water and wastewater exceeded 5% of the GDP in 2004, more than double the recommended percentage. Unless the GDP is more than doubled, Treasury subsidies are inevitable if the contribution of beneficiaries to water and wastewater cost of water services is to cover only 2% of the GDP. Higher contribution will tax other components of the family budget, affecting the quality of life.

Because of the difference between water tariffs and cost of service provision, the WAJ has been incurring net financial losses every year since 1985. However, since 1998, in large part because of upward water tariff adjustment, the WAJ was able to approach the recovery of operating expenses only. The upward adjustment of tariff drew public uproar at the time and was used for political ends by parties opposed to the peace treaty with Israel. In advocating the peace treaty, the government cited water gains as one of the benefits the treaty brought to Jordan. How is it, the opposition asked, that water becomes more expensive when the supply is increased by the treaty? There are no gains from the treaty, they concluded, and the disenchanted beneficiaries concurred.

More upward adjustments are needed in the water tariffs for both municipal and irrigation water to approach cost recovery. In this regard, the government faces a dilemma: on the one hand, budget deficits are to be controlled within permissible limits; on the other, the adjusted tariff will cause the expenditure of higher percentages of the GDP on water and wastewater services, a reason for public protest and more political hardships. The average supply per capita is about 46 m^3, and little room exists for reducing the supply rate and saving on expenditure.

Some argue that when a family is out of network water, it purchases from water vendors and pays as much as US\$2.82/m^3, double the highest rate in the water tariff. Although this is true, it is not an everyday event, and the high cost prompts poorer families to cut back on hygienic uses of water and limit them to drinking and cooking.

Sector Reform Initiatives

For a long time, as is the case in many developing and even developed countries, Jordan's water management was oriented more toward supply than demand.[3] Water administrators and managers focused on the development of new groundwater and surface-water resources to boost the supply. However, in the 1980s, the cost of developing new water resources became progressively more expensive. Deeper wells had to be drilled in more remote areas, and more expensive dams had to be built to regulate modest and highly fluctuating flows. Carryover storage became essential, demanding more storage volume than annual flow regulation would need.[4] Water had to be transported from remote locations with high total dynamic heads, imposing higher marginal costs of delivery and distribution. Moreover, the costs of treatment and disposal of wastewater also became increasingly expensive as new wastewater treatment plants were erected and older ones had to be expanded and upgraded to handle the increased flow of wastewater.

This situation was aggravated by the Gulf War in 1990, when the country witnessed a 10% increase in its population over a short period of time; the newcomers who had lost their jobs in the Gulf and did not want to stay there resettled in the Amman–Zarqa area, which feeds the As-Samra wastewater treatment plant. With little preparation for the utility needs of the newcomers, the water and wastewater systems came under severe stress. This paralleled a severe economic and fiscal crisis the country was going through.

Demand Management

Under the above circumstances, the government of Jordan in the late 1990s shifted the focus of water policy from supply management to the less extravagant but difficult emphasis on demand management. Difficult as it was, the new alternative was much cheaper and wiser. Supply management could not be avoided, but the shift of focus to demand management enabled water managers to save water and buy time. It became evident that the groundwater renewable aquifers could not be called upon for more supplies, as they had become overstressed, and reliance had to extend to fossil freshwater despite the environmental risks involved. Reliance for supply also shifted to more surface water from the Yarmouk River, the source that was the dream of Jordan Valley developers for half a century.[5] In fact, a progressive decline in the supply from the Yarmouk since 1970 had prompted the Jordan Valley farmers to adopt advanced irrigation methods and cut down water consumption.

The government then moved to raise the irrigation water tariff fivefold on average, from US$0.004/m^3 in 1973 to an average of US$0.021/m^3 in 1996. This is still far below the cost of operating the irrigation system. It also adjusted the municipal water tariff upward, and the WAJ revenues now approached the recovery of the operating expenses.

Along with improved tariff structures, the government embarked on an important program for rehabilitation of the distribution systems and reorganization of municipal water service zones inside cities. The replacement and rehabilitation of old networks in Amman helped reduce leakage, and the reorganization of service zones helped improve water service, using gravity to carry water from storage tanks to the beneficiaries instead of direct pumping from the source to the distribution network.

Reforming the Water Sector Policy

Under the water strategy, the Council of Ministers approved a set of water policy papers: Water Utility Policy, Irrigation Water Policy, Groundwater Management Policy, and Wastewater Management Policy. The strategy, supplemented by these policy papers, set objectives for resource development, resource management, legislation and institutional setup, shared water resources, public awareness, performance, health standards, private sector participation, financing, and research and development. A water sector investment plan covering the period 1998 to 2010 was also prepared and presented to the donors at a conference in Petra in September 1997.

The most recent statements of government policy with respect to the water sector are the Water Sector Action Plan, adopted by the Council of Ministers in December 2002, and the Water Sector Plan and Associated Investment Program 2002–2011, an update of the 1997 program. The Water Sector Action Plan set out specific steps for implementation in the institutional and legal context, with policies on agricultural water use, cost recovery, private sector participation, information systems, and adoption of international conventions on biodiversity and desertification. Key measures included the following:

- establishing companies under Jordanian law, owned by the WAJ, to provide water services on a commercial basis in Amman and Aqaba as of 2004;
- implementing water loss reduction programs in Irbid and Jerash, and beginning a similar program in 2006 in Karak;
- increasing water tariffs by 15% in 2005;
- reducing the number of WAJ employees by 1% annually; and
- achieving O&M cost recovery targets of 147% in 2003, 139% in 2004, 135% in 2005, and 148% in 2006.

These represented the most advanced policy statements ever made by the government. The shift from government provision of water and wastewater service to entrust the same to companies was a bold step in the right direction.

Implementation of this policy statement took place in Aqaba in 2003 by establishing the Aqaba Water Company, with shares taken up, as a first step toward privatization of services, by the WAJ, Aqaba Special Economic Zone Authority, and Social Security Corporation. The Aqaba Water Company was pronounced operational in August 2004. A summary of the key elements of these policy documents follows.

Quality and Service Levels. The policy provides for the securing of water services at affordable prices and acceptable standards and extending services to remote and less developed areas. The most important parameters for service level assessment include the following:

- maintainance of water quality;
- frequency of summer water supply;
- frequency of winter water supply;
- response time for repair of network leakages, pressure loss, and sewer blockage;
- reduction in waiting times for water and wastewater connections; and
- reduction in waiting times for resolution of customer complaints.

Cost Recovery and Pricing. Recovery of operating and maintenance costs is to be targeted as a standard practice, with the ultimate aim of full cost recovery. Water tariffs are to be reviewed and adjusted regularly, based on costs of supply, operations, and data analysis. In addition, the movement toward full cost recovery is to be linked to average GDP per capita as well as the cost of living and consumer basket indices. Profitable undertakings in industry, tourism, commerce, and agriculture, however, are to pay a fair water cost. Differential prices for water are to be adopted based on quality, end users, and social and economic impact in different sectors and regions.

The affordability issue and the full cost recovery target are, in the case of Jordan, contradictory. Slowdowns of economic growth and a sharp increase in oil prices have been major factors that hampered the achievement of higher percentages of cost recovery. This is a dilemma in search of a solution, and it can be resolved through comprehensive social, economic, and political reforms. A genuine effort, led by the king, has been under way to eradicate corruption, reduce undue public reliance on the government for employment and services, assure equal opportunity to all Jordanians, eliminate overemployment in the government, encourage public participation, and reform election law. Many other measures will be detailed in a national agenda that is now being worked out.

The above measures are hoped to increase efficiency of performance, speed up the economic and social development pace, and optimize the cost of service provision, leading to better affordability by the beneficiaries.

Water Reallocation. Water reallocation is going to be a formidable task. Farmers using groundwater will not give up their resources easily to have them transferred to municipal use. Those using fresh water will not want to exchange it for gray water. Some incentives have to be built in, such as the purchase of wells, changing the zoning of land use, compensations, and the like.

Enforcing Groundwater Regulation Number 85 for the Year 2002. Groundwater Regulation (Bylaw) Number 85 for the year 2002 has all the good intentions of curtailing overpumping from groundwater aquifers in conformity with the Groundwater Management Policy and Jordan's Water Strategy. This bylaw sets the rules for licensing the drilling of wells and water extraction for agricultural, municipal, and industrial uses. It also delineates the conditions that need to be met for a permit to be issued, setting the appropriate distance from other wells and springs and requiring a pumping test, among others. It also sets the rules for the licensing of drilling rigs and drillers, with the law prohibiting any person or legal entity from obtaining or using a drilling machine or practicing water drilling work without obtaining the appropriate license from the WAJ.

What created a widespread buzz in Jordan about this bylaw, however, was that it sets charges for the overabstraction of groundwater resources for irrigation. The charges are variable and depend on the user sector, the quantity abstracted, and the condition of the aquifer from which the well draws water. They differentiate between licensed and unlicensed wells, also saying that owners of unlicensed wells should have their status rectified by applying for a license and shall pay an overcharge. Industrial well users pay higher tariffs than agricultural well users, and users in the Azraq area pay higher tariffs than users in other areas.

The government's success in achieving these policy objectives has been modest. It faced considerable resistance to its attempts to confiscate unlicensed drilling rigs and prevent unauthorized drilling of wells, especially in the areas north of the Dead Sea. Some of its officers in charge of supervision were faced with armed resistance by natives seeking to drill unauthorized wells. Success in enforcement of abstraction rates to achieve balance between inflows and outflows of aquifers was, despite all the good intentions, very modest. In trying to curtail abstraction from the Disi aquifer, the WAJ encountered resistance from farming companies that had rental contracts with the government; these specified land rentals but not water charges. The dispute with the companies went through litigation in court, which ruled that the WAJ was collecting the surcharge legally in accordance with an effective bylaw. However, in August 2005, the new government decided to exempt the farming companies in Disi from paying the surcharge on abstracted water because of the lack of any stipulation of a water tariff in the contracts.

The government's enacting of the bylaw that empowered the WAJ to collect a surcharge for water abstracted from any well in excess of a specified free quantity has the following shortcomings:

- Constitutionally, the WAJ does not represent the public that owns the water; rather, the Treasury does. Any funds collected as surcharges for water abstraction should be done for and on behalf of the Treasury, not for a WAJ fund that is, by its law, independent of the Treasury. Furthermore, any collection of funds should be stipulated by law. A law, and not a bylaw, should be enacted for collection of the surcharge funds, which should be deposited in the Treasury.
- Also constitutionally, charges cannot be levied from citizens unless Parliament passes a law according to which such charges can be levied. A bylaw is passed only by the Council of Ministers. Any such regulation has to be in conformity with the law under which it is issued, in this case the WAJ Law of 1988 and its amendment. But that law empowers the WAJ to collect charges, levies, and tariffs against services rendered by the WAJ to the beneficiaries. In the case of groundwater abstraction, the WAJ provides no service against which its law can be applied.
- Legally, the WAJ or the Treasury is entitled to charge for water in exchange for an effort that either has exerted to make the water available. In Islamic Sharia, a source of legislation in Jordan, "water is not sold unless an effort is made to make it available" (Abu Yusif d. 804, 97). Although the WAJ is thus entitled to charge for water it provides by virtue of effort it has put into building networks, pumping water, treatment, and so on, neither the WAJ nor any government entity has contributed to the abstraction of water from private wells, and therefore, no charge on any quantity of flow is warranted. As an alternative to the provision of a surcharge, a penalty can be imposed on any well owner who exceeds the abstraction rate specified in the abstraction permit, and that penalty can be financial. If repeated, this can result in a prison term or closure of the well. A measure of this type would be compatible with the provision of the Sharia and would accomplish the intended goal.

Future Trends in Water Tariffs

The future will bring further increases in the tariffs as per Jordan's 2002 Water Action Plan. The government expressed its intention to continue with its aim to achieve cost recovery, with the goal of attaining 135% of the operating and maintenance costs in 2005 and 148% in 2006. Plans to launch several major infrastructure projects will necessitate that tariffs be raised in order to recover operating and maintenance costs as well as a portion of the capital costs, as per Jordan's Water Strategy. A study made by the Ministry of Water and Irrigation expected that water tariffs would be increased by 8% in 2005 and a further 8% in 2006. However, it it is difficult for the government to enforce an upward water tariff adjustment, given the high rise of fossil oil cost in 2005. On September 20, 2005, the government approved a rise in the prices of refined oil products in percentages ranging between 5% and 20%. That was the second increase in fuel prices in two months. More than 65% of the customers pay less than US$10 per quarter—which is less than US$0.15 a day for a family of five or six. Jordan has more than 3 million subscribers of cell phones with an average individual bill of US$1 a day.

It appears as though the goal of achieving cost recovery could be elusive, as the exogenous factors of cost are not under government control, and the ever-increasing oil prices demand higher electricity rates and thereby higher water tariffs. Oil grants from Iraq came to an end in 2003, and the replacements from Saudi Arabia and Kuwait stopped in 2005. Jordan has to cope with the free-market prices of oil and manage its affairs accordingly. Only increased efficiency, transparency, stakeholder participation, and competent management can bring Jordan closer to cost recovery.

Water Conservation Trends

Jordan's Water Strategy and its supporting policies emphasize the need for water conservation, which is to be achieved through a mix of instruments, including regulation, system efficiency improvements, public awareness, and economic incentives. Significant stress for water conservation measures is placed on the agricultural sector. Not surprisingly, a 5% improvement in irrigation efficiency results in the saving of 26.1 mcm of water in 2002 figures of consumption. Such a saving could be used to reduce overabstraction of groundwater or allocated to municipal uses, where the saved water could be sufficient for half a million people that year.

In terms of actual measures implemented to achieve water conservation, the MWI, assisted by the United States Agency for International Development (USAID), initiated a project called Water Efficiency and Public Information for Action (WEPIA) and is continuing its implementation in the form of a sustained public awareness campaign on water conservation. Moreover, significant improvements in equipping private wells with water meters have been achieved, and meters are now installed on 93% of all wells within the kingdom. The MWI also has been closing some unlicensed wells and impounding illegal drilling rigs. There is practically no waste in treated wastewater, as it is all reused in agriculture. The JVA, supported technically and financially from Kredi-tanstalt für Wiederaufbau (KFW), is formulating an action plan for reuse of treated wastewater in agriculture, drawing on lessons from an ongoing pilot project. The Arab Fund for Economic and Social Development extended support in 2004 to the National Center for Agricultural Research and Technology

Transfer of the Ministry of Agriculture to conduct research on crops irrigated with treated wastewater.

Private Sector Participation

In recent years, the government has entered into partnership with the private sector for the management of water and wastewater. Several of these arrangements are examined in this section.

Amman Water and Wastewater Management Contract

On July 31, 1999, the government of Jordan signed a contract, the first of its kind in the country, for the management of Amman water and wastewater facilities. The design of the contract was done with technical assistance from the World Bank, and it was similar in many ways to contracts the bank helped design in Gaza and Trinidad. The consortium that was awarded the contract became responsible for managing all water and wastewater operations within the Greater Amman area.[6] Capital investment and the upgrade of facilities remained the responsibilities of the WAJ. The remuneration mechanism under the contract consisted of a fixed fee of US$8.8 million for four years, and an annual incentive fee equal to 5% of the incremental cash flow was made a function of increased revenue credited to improved performance of operations. The management contract for Amman was extended for two and half years and will come to an end on December 31, 2006.

Shortcomings in performance of the contract were attributed to delays in the capital investment program of the WAJ, the reluctance of the WAJ to accept the return to it of more employees as per the contract, and droughts that caused water supplies to fall short of the levels envisaged in the plans. However, many advantages were cited, including technology transfer and introduction of "best practice" in the management of facilities, upgrades in customer relations and communications, improvements in the environmental and public health monitoring program, increased revenues, reductions in unaccounted-for water, more advanced meter-reading systems, and improved response to emergencies. After all, rules and regulations governing the consortium are different from the government rules and regulations that the WAJ had to observe. The financing of the operating and maintenance costs also was superior to the allocations made under the WAJ budget before the management contract was introduced, thanks to loans extended by the World Bank.

It is believed that the WAJ can go further in shifting toward private sector orientation; the question is how to operate in a private sector mode with government funds. Government subsidies to the WAJ will have to keep financing capital investment, and the government may also have to continue to help meet the escalating cost of operations in light of the sharp hike in the cost of energy.

Still, the PMU established in the WAJ could be made more dynamic by assigning it the proper authority for performance, overhauling the procurement regulations for better efficiency and faster response time, and staffing the authority with new recruits on the basis of qualifications and competence.

Aqaba Water Company

For many years, the Aqaba Governorate Water Directorate (AGWD) of the WAJ functioned with positive operating cash flow and contributed to the treasury of the WAJ, which helped subsidize other WAJ directorates elsewhere. This was due to favorable supply conditions, with good water availability from the Disi–Aqaba pipeline, and the composition of its customers, with more than 50% of water consumed by industries and tourism facilities, at a rate of US$1.41/m^3 for water and US$0.71/m^3 for wastewater.

Moreover, the AGWD received substantial investments from funding agencies to upgrade its water and wastewater systems, with minimal requirements of capital investments in the years to come. The loans were mostly absorbed by the government in 1998 as capital subsidies to the WAJ and added to the government's debt burden.

The government, through legislation in February 2001, transformed Aqaba into a special economic zone. The master development plan for the Aqaba Special Economic Zone (ASEZ) forecast ambitious investment by 2020 of about US$6 billion in tourism, industry, and services. The plan predicts a population level in the ASEZ approaching 280,000 people by that year, currently estimated at around 110,000 for the governorate (DOS 2005).

This prompted the MWI to consider a commercialization program for Aqaba water and wastewater, which allows operations under private laws of Jordan. The program started with the establishment of the Aqaba Water Company, announced operational on August 1, 2004. The company is now owned by the WAJ (85%) and ASEZA (15%) and is empowered to operate on a commercial basis. The autonomous water utility operates with market rules, and various operational improvements are noted by the residents of Aqaba. The effluent of the wastewater plant there will be sold to the phosphate industry.

As-Samra BOT Contract

The As-Samra wastewater stabilization ponds were constructed in 1985, with a design capacity of 68,000 m^3/day. The plant receives a daily flow of 170,000 m^3 (as of 2004) and is overloaded biologically and hydraulically. The plant serves about 2.2 million people, 45% of Jordan's population. Because of the serious environmental problems caused by the plant in the Amman-Zarqa

Basin, the MWI decided to change the treatment technology and operational arrangements.

In 1999, a build-operate-transfer arrangement was considered by the MWI in cooperation with USAID, based on a study of different investment scenarios and a review of feasibility and technical designs. After a few years of procurement and contract negotiations, the first BOT contract in the water sector was signed in December 2003. The contract extends over 25 years and includes construction of an activated sludge mechanical treatment plant at As-Samra and expansion of the Ain Ghazal pretreatment plant and pumping stations, the first treatment plant ever built in the country. The new treatment plant will have a capacity of about 270,000 m^3/day and will serve the populations of Greater Amman, Russeifa, Zarqa, and Hashimiyya.

USAID provided US$78 million for plant construction, and the government of Jordan contributed about US$14 million for social, economic, and environmental considerations, together about 50% of the project's capital. Of the remainder, 20% came from the implementing consortium's own equity and 30% from a consortium of banks led by the Arab Bank as a loan to the implementing consortium.[7] A treatment fee of about US$0.20/m^3 will be paid by the WAJ to the implementing consortium when the plant starts operation.

To achieve cost recovery, the government of Jordan raised wastewater tariffs in Amman and Zarqa by about 9% in 2001. The treatment charges are sufficient to meet debt service and fixed operating expenses agreed on in advance and are payable monthly throughout the operating period. The project will achieve substantial environmental and economic benefits and provides for capital investment and participation of the private sector in wastewater treatment.

Northern Governorates Internal Competitiveness Program

As part of the government policy, the MWI and WAJ have been considering options for water and wastewater services for the northern governorates of Jordan: Irbid, Jerash, Ajloun, and Mafraq. Supported by KFW, a fixed-term management contract was considered by the MWI, a request for proposal was prepared, but the procurement process has failed to attract a management contractor for the northern governorates.

With assistance from KFW, the government is looking at other options, including an Internal Competitiveness Program (ICP) that will be implemented by the Northern Governorates Water Directorate (NGWD). The ICP concept is widely known in the United States and other countries and is used primarily to increase operational efficiency and improve services. The ICP initiated in the United States in the 1980s to avert the privatization of public utilities served as a guide. In simple words, the public administration has to match private sector performance and efficiency so that the management will not go to the more efficient private sector.

The main features of this program for the NGWD are similar to the Amman management contract, with more autonomy for the operator and delegation of authority by the WAJ secretary general to the NGWD in connection with staffing policy, budgeting, and investment in maintenance and repair programs. The ICP will allow more flexibility in procurement, recruitment outside civil service regulations, and use of local and international consultants to assist NGWD management.

Water Users Associations and Farmers Committees

Recent experiences have revealed that the participation of farmers in managing part of the irrigation system usually results in achieving a higher level of efficiency in water use. Yet in order to effectively participate, farmers need to be organized in larger bodies, often known as Water Users Associations or Farmers Committees.

In view of the JVA's continual efforts toward achieving optimum and efficient use of available water resources, a project was adopted to enhance farmers' participation in managing the distribution of irrigation water in several districts in the Jordan Valley and Southern Ghors, as well as in helping identify and resolve the technical and administrative issues related to the water distribution system. In planning the project, situational analysis was done to understand the problems of both the JVA and the farmers. The project's main objective was to enhance transparency and effectiveness in water use and distribution. Other objectives included increased water use efficiency, decreased costs, reducing JVA staffing needs (through farmers' participation in water distribution and maintenance), creating trust between the JVA and the farmers and among the farmers, preventing illegal practices, and enhancing farmers' satisfaction and care for the infrastructure.

The first phase of the project was completed by the end of 2003. It set the outline for farmers' participation, promoted the benefits of farmers' committees and associations, and identified the roles and responsibilities of such groups. About nine operative associations or committees currently exist in the Jordan Valley. Of these, two are Water Users Associations, officially registered as cooperative societies, and the rest are Farmers Committees, which are internally approved to perform the duties of Water Associations. The main responsibilities of Water Users Associations and Farmers Committees in the Jordan Valley are as follows:

- representing farmers before the Jordan Valley Authority (reporting issues, suggestions, and complaints);
- managing irrigation lines (monitoring, reporting deficits, performing minor maintenance activities);
- locating and reporting system leakages;

- opening and closing water intakes;
- controlling and reporting water theft; and
- controlling tampering and vandalism.

At a later stage, these associations and committees will be involved in preparation of irrigation scheduling, as is currently the case in Ghor Al-Mazra and Haditha of the Southern Ghors. Joint water associations also may participate in water allocation issues at the level of the entire valley.

The experience of such associations in Jordan is growing, and their benefits already are appreciated by the farmers and the JVA alike. Where associations are functional, the number of complaints by downstream users has been reduced substantially, and the amount of damage and vandalism to the JVA system has decreased. Farmers receive water according to schedules with appropriate pressure for their systems. The level of confidence and trust between JVA and the farmers has increased and is being reinforced.

Looking Ahead

Jordan's water stress cannot be alleviated without a dual approach of demand and supply management. The future course of action is briefly described in this section.

Demand Management

Demand management has to focus on upgrading water management capabilities through human resource development, the adoption of automation in water management, and continually increasing water use efficiency. This would lead to a sizable reduction in manpower currently on the job. The focus on Water Users Associations will help control the cost of water to users, reduce unaccounted-for water, and improve water use efficiency. The associations will contribute to ensuring equity in distribution. They are also a practical way of involving the stakeholders and assuring transparency.

Public awareness campaigns to conserve water have been under way and should intensify in the future. They will enhance efforts to improve water use efficiency and can pave the way to the gradual adjustment of water tariffs. Transparency of water management is essential to gain public support. The efforts aiming at reform and combating corruption will help create an environment of trust and partnership between the public and the government.

The issue of water tariffs will loom over the water sector for a long time to come. A discrepancy exists between water cost and the average share of per capita income. Cross-subsidies are built into the current water tariffs but are not sufficient to assure full cost recovery. The increase of water consumption has not been natural as a result of waves of refugees, displaced persons, and returnees; an influx of foreign laborers from Egypt, Iraq, and Syria; and the refuge Jordan has been giving to Iraqi families escaping the violence that has been plaguing their country since Operation Iraqi Freedom led by the United States in March 2003.

The cost of living has increased, particularly for imported commodities. Jordan imports about 75% on average of its food commodities, all of its fossil fuel needs, and a long list of manufactured commodities needed for production and services. The GDP did not grow in parallel with the increased cost of living, and the adjustment of water tariffs, although essential to maintain good service, will face difficulties. An accelerated growth in the GDP is needed to meet the cost of living; failing that, the quality of life will regress, and it is the poor who will suffer most.

Supply Management

Because of the increased demand for municipal, industrial, and irrigation water, along with the reduction in Jordan's water shares in the Yarmouk as a result of increased Syrian withdrawals, most of Jordan's renewable water resources have been put to beneficial use.

The bilateral agreement between Jordan and Syria over the Yarmouk should be revisited to address issues basic to water sharing. Groundwater abstraction in Syrian territories of the Yarmouk Basin was overlooked in both the 1953 and 1987 agreements. Syrian abstraction of groundwater from the aquifer that feeds the lower springs (below elevation 250 m above sea level) has drastically reduced the base flow of the river at Jordan's expense. Additionally, coordination in the filling of surface reservoirs in Syria and the Wehda Dam reservoir in Jordan is essential to ensure the returns from all these dams. Syrian dams have to be checked against the number allowed in the 1987 agreement.

Supply management will have to depend on new water sources. The ones in the pipeline at this time are the Red Sea–Dead Sea linkage and the exploitation of fossil freshwater resources. The Red Sea–Dead Sea linkage will use the net potential energy resulting from the difference in elevation between the two seas—about 418.30 m in 2006—to desalinate seawater and generate some electrical energy. Linking the two seas will be a combination of piped conduit, to lift the water from Aqaba and clear the Gharandal Divide in Wadi Araba at elevation 250 m above sea level, and open canals and tunnels traversing Wadi Araba. Energy will be needed to pump water in the piped section, and the rest of the flow will be by gravity. The project is capable of producing about 500 mcm of fresh water at steady state (more can be generated in the initial 10 years to bring the Dead Sea level up to –396 m, or about 21 m additional depth of water). The resulting brine and the residual drop will be used to operate generators and produce electricity needed to pump the fresh water to the places of use.

The benefits and costs of this megaproject have to be shared with Israel and the Palestinian Authority (future state), as both are riparian parties on the Dead Sea. Jordan's share of its surface area is about 50%, while Israel and the Palestinian Authority share the remaining 50%. Massive foreign assistance is needed to implement this project, a cooperative effort between former enemies that helps stabilize peace.

Utilization of fresh fossil aquifers is the more palatable option considered by the government at this time (2006), focusing on the remote area of Disi in the southeast of the country. A project calling for the abstraction of 100 mcm/y has been studied, designed, and floated in a BOT tender, but financial considerations prompted the government to cancel the tender after bids were analyzed. Transparency was lacking in the management of this crucial undertaking, and the media reported the cancellation of bids and the award of the project to a company owned by the armed forces. But that, too, was aborted for reasons unknown to the public.

The implementation of the Disi–Amman municipal water project was decided more than a decade ago, before proper investigation of the extent of the fresh water in the sandstone layer of which the Disi aquifer is part. In 1999, an exploratory well drilled in the Lajjoun area east of the city of Karak yielded fresh water from that sandstone layer. More wells were drilled in its proximity to supply Karak and Amman with fresh fossil water. More exploratory wells yielded encouraging results south of Lajjoun at Hasa and at the phosphate mines east of Hasa.

It would be prudent on the part of the Ministry of Water and Irrigation to conduct an exploratory program on the sandstone layer that extends under practically the entire kingdom before huge investments are made to transport the Disi water to Amman. It is likely that closer well fields could be drilled that would require much less capital to transport their water to Amman and other cities.

Sustainability and Affordability

The sandstone aquifer containing the fossil water under Jordanian territories is huge and extensive. The Disi part alone is estimated to sustain the Disi–Amman municipal water project's capacity (100 mcm/y) for about 150 years. It is recognized that fossil water is not renewable, but it is also recognized that fossil fuel, the source of most of the world's energy supplies, is not renewable either.

Seawater desalination is an option for the generation of water for municipal and industrial purposes. The process is energy intensive, and today's technology is said to enable the generation of sweet water at a cost of about US$0.50/m^3 at the seashore. This needs to be updated in light of the recent hikes in fossil fuel prices; it will cost more to transport and distribute the produced water. The closest Jordanian sea coast is Aqaba, though it is far from the main cities, which are also high in elevation. The cost of delivery of desalinated water to Amman from Aqaba exceeds US$3/m^3, too high to meet, especially when the costs of distribution and unaccounted-for water are added.

Proponents of fossil-water use recognize the environmental impact inherent in such use, but they point to the eventuality of the invention of an environmentally friendly energy-generation technology to save the world from virtual collapse when the fossil fuel reserves run out with no alternative in sight. The momentum of human civilization will pave the way for new technologies in energy generation. They point to the historic sequence of energy generation, from wood to coal to fossil fuel to nuclear fission, and say that the time is coming for the next quantum jump in the technology of energy generation. When this happens, it will be possible to desalinate seawater and pump it for human use. With this eventuality, they continue, it becomes a matter of holding out: could the fossil-water reserves outlast the fossil fuel reserves?

It looks as if Jordan's fossil-water reserves will outlast what is known of energy reserves in the world. Even if the quality may sometimes be brackish, it would be much less expensive to desalinate than seawater. As a matter of fact, Jordan has embarked on the desalination of brackish water already.

Both supply management options are clouded by the cost of supplies as compared with the per capita share of the GDP. Both are expensive, and subsidies are inevitable. Although the Red Sea–Dead Sea linkage may receive international support for its role in stabilizing the Middle East, the use of fossil water has not attracted assistance from friendly countries because of its environmental impact and the fact that the limited, nonrenewable supplies will eventually run out.

The answer to the issue of affordability is a dual approach of demand management and accelerated economic growth. Business as usual in economic growth rates is not sufficient to cope with the cost of incremental supplies in the absence of Treasury subsidies.

Diversion of Agricultural Water to Municipal Use

A faster and ready option for supply management of municipal water is to divert the groundwater now being used in irrigation to municipal uses. Diverted water can be replaced with gray water from wastewater treatment plants nearby or with shadow water if needed (Chapter 7). Groundwater used in irrigation in the plateau region is close to population centers where it is needed for municipal uses.

The only factor that raises concern is the social and environmental impacts of such diversion. A careful and thorough social survey should be made to identify the jobs that would be lost and the current beneficiaries who would seek alternative jobs. It is not a secret that most agricultural labor jobs are taken by non-Jordanian workers, especially on the plateau, where groundwater is being used in agriculture.

Diversion of surface water to municipal uses has been practiced since 1986, when the Deir Alla–Amman pipeline was put in operation to pump water from the King Abdallah Canal to Amman. The capacity of the pipeline was doubled in 2004, and water transferred from Israel under the peace treaty in addition to desalted brackish water is being pumped to Amman. More surface water has been reallocated from agriculture to municipal and industrial uses from the Mujib flows and the flows of springs discharging into the Dead Sea from the eastern escarpment.

Management of Gray Water

Treated wastewater has become a primary water resource, especially for the middle Jordan Valley. Gray water from Amman and environs compensates for the surface water diverted to municipal and industrial uses in Amman. More emphasis will be placed on treated wastewater for irrigation use in Jordan. Flows will increase with the increase of municipal supplies. More practical research will be done on gray-water use, especially at the National Center for Agricultural Research and Technology Transfer and at schools of agriculture in the Jordanian universities.

Management of Green Water

Green water in Jordan is about equal to the blue-water renewable resources, and it is used in rain-fed agriculture. The utilization of rain-fed lands has not been optimum because of uncertainties of the rainy season and other social factors. Leaving rain-fed lands fallow in any given year is tantamount to losing green water stored in their soils that year to evaporation. Because water is public property, landowners should be made aware of this, and the fact that wasting public property is prohibited by law. The need for future legislation to address this and other issues is discussed in Chapter 2.

Summary

Jordan's water resources are under excessive stress, with 2,717 persons per unit flow of water in 2004 compared with 588 persons at equilibrium conditions. Accordingly, the country is subject to severe water strains and faces formidable challenges. At the heart of such challenges is how to optimize social, economic, and environmental returns from the scarce water resources. Energy shortage exacerbates the costs of water stress, because water service is invariably associated with energy expenditure.

The most difficult challenge is water governance. Water administration, after vacillating between multiple agencies and central agencies, finally embarked on a central administration with a cabinet portfolio. Shortcomings under the central administration of the Ministry of Water and Irrigation are enumerated, and they can be rectified. Responsibilities are shared with the Ministry of Health (drinkable water) and Ministry of Environment (water quality).

Stakeholder participation in water decisions was initiated in the 1960s through consultations with donor agencies, primarily the IDA of the World Bank. Participation was reflected in the membership of the Board of the Jordan Valley Authority and its farmers' selection committee. Furthermore, stakeholders' partnership with government was best reflected in their care for on-farm investments and production, manifested in the government-sponsored irrigation infrastructure in the Jordan Valley and in irrigated farming elsewhere.

Other aspects of governance include social equity and enabling the poor. Water rationing and tariffs, as determined by the government, reflected its desire to address social equity and subsidize water consumption by the poor.

Challenges facing the water sector include securing incremental supplies, unsustainable uses of groundwater, water allocation and reallocation, and human resource development. The exogenous factors of the costs of water services, such as energy costs for operation and technological hardware, are challenges that the government has to face and manage. Public-private partnership in water resource management has begun and is being promoted.

A formidable challenge facing the water sector is cost recovery. This is linked to the GDP and the willingness of beneficiaries to pay. The costs of water and wastewater service provision have been increasing faster than the economic growth rate, but accelerated economic development is necessary in order for beneficiaries to be able to afford the bills for the actual costs. Nonetheless, the irrigation water tariff was raised in 1995, followed in 1997 by a municipal water tariff adjustment, thus reducing the government subsidy to both.

Like most developing countries of the world, Jordan focused on water supply management for a long time before circumstances of demand, supply, and financial capacity forced a new focus on demand management. The government issued a water strategy and a set of policies in 1997–1998, and it subsequently set forth an action plan to improve the quality of service, address the cost recovery, partner with the private sector, allocate water, and control groundwater abstraction.

A contract for the management of water and wastewater in the Greater Amman area has been in operation since 1999 and is due to come to an end in 2006. A BOT contract was let for the upgrading and expansion of the As-Samra wastewater treatment plant. A company has been established in the Aqaba Special Economic Zone to take over the responsibilities of the WAJ there. A Water Users Association has been set up in the Jordan Valley.

Finally, predictions are made of what may be expected in the future in terms of both supply and demand management, reallocation of blue water, and the management of green water and wastewater resources.

References

Abu Yusif Ya'koub Bin Ibrahim Bin Habib Al Ansari. Died 804 CE. *Kitab Al Kharaj,* section on canals, wells, and rivers, 94–97. Beirut, Lebanon: Dar Al Ma'rifah for Printing and Publishing. Reprinted from an original manuscript, no. 674, and verified with the copy printed by Boulaq Press, Egypt, 1924.

DOS (Department of Statistics). 2005. *Jordan in Figures: 2004.* Amman, Jordan: DOS.

Notes

1. The chronic resource problem in Jordan is summarized in one word, *liquidity:* water, oil, and funds.

2. An irrigation lateral refers to a branch of a main canal conveying water to a farm.

3. A wave of grand projects of planning and implementation spread from the United States to the rest of the world. Dams were promoted and built, grand irrigation schemes were implemented, and interbasin transfers became common practice. Examples in the United States exist in the Columbia Basin, on the Colorado, and in Southern California; examples elsewhere can be seen in the dams of Turkey, in the Indus Basin, and most recently on the Yellow River in China.

4. Examples are the Mujib Dam, Wala Dam, and Tannur Dam.

5. The Wehda Dam, the predecessor of the Maqarin Dam, was first embarked upon in 1952 for the regulation of the Yarmouk River to irrigate the Jordan Valley. Recent water allocation assigns a good part of its water for municipal uses in the cities of North Jordan.

6. The consortium consists of the French Lyonnaise de Eaux, the British Company Montgomery Watson, and the Jordan Company Arabtech Jardaneh.

7. The implementing consortium consists of Suez Environment (20%); Ondeo Degremont, Inc. (30%); and Morganti Group (50%).

9

Linkages with Social and Cultural Issues

Munther J. Haddadin and Musa Shteiwi

The people of Jordan are of several ethnic and national origins and religious persuasions. The majority consists of Arab Muslims, originating in Jordan and the Arab countries around it; the remainder are indigenous Arab Christians, deeply rooted in the country, and non-Arab Muslims, originating in the Caucasus and Kurdistan. Jordanians' cultural affiliation is Arab Islamic, molded by Arab and Muslim scholars and thinkers over 14 centuries since the emergence of Islam in the early seventh century.

No being or species can exist without water, but to Muslims and Christians, water has earned a religious significance as well. A person is baptized into the Christian faith with water, and an infant is bathed immediately after being pronounced Muslim at birth. Water has many spiritual meanings and rituals in both (and other) religions. Several verses in the Holy Qur'an speak of God's blessings in making water available to mankind and to land. The Prophet Mohammad is quoted to have said, "People are partners in three: water, pastures, and fire" (Abu Yusif d 804, 96). The importance of water as stipulated in religion is basic in shaping the attitudes of people toward it, so much so that many uninformed people have ventured to say that Islam has decreed that water is free and no charges can rightfully be asked for it. However, the prophet's wife, Ayisha (Um Al Mu'mineen, or mother of all believers), was quoted as saying that she heard the prophet prohibiting the sale of water. Islamic scholars interpreted this to mean that water may not be sold unless an effort has been made to obtain it (Abu Yusif d 804, 97); in other words, water at its natural location of occurrence cannot be sold, but if a "water man" delivers it to the user, he can charge for the service. The water can be conveyed only

211

through delivery means, not through wells or basins (Abu Yusif d 804, *97*). Today's delivery instruments are piped networks and pumps or water tanks.

The Muslim judge Abu Yusif Ya'koub Bin Ibrahim decreed that whoever has a spring, well, or canal does not have the right to prevent passers-by from drinking or watering their animals from it. The owner does not have the right to sell any of it for drinking purposes, for either humans or animals. The owner does, however, have the right to prevent irrigation of land, plants, palms, and trees, and no one has the right to use someone's water to irrigate any of those without the owner's permission. Even then, however, the owner does not have the right to sell the water (Abu Yusif d 804, *95*). The essence in the Islamic tradition thus is that water can be sold only if an effort has been made to obtain or serve it; in its natural occurrence, water can be sold neither for drinking nor for irrigation.

The principle of equity is embedded in the Islamic tradition. A clear example is evident when prayers are performed five times a day: all worshipers, rich and poor, young and old, stand side by side in alignment, led by the imam, who stands in front. The hadith that says that Muslims are partners in water, fire, and pasture entails social equity in benefiting from water, with the consent of the owner, to the degree of partnership.

The treatment and reuse of wastewater increase the benefits from the unit flow of freshwater resources, thereby increasing the efficiency of use. Several uninformed people recommended disposing of municipal wastewater away from land application, on the grounds of public health concerns and fear of contaminating the soil with the microbes the wastewater carries. Because of the aridity of most of the Muslim countries and the need to augment the freshwater resources, the necessity arose for an Islamic fatwa to clarify the position on the reuse of treated wastewater. The Commission of Grand Scholars in the Kingdom of Saudi Arabia, in its 13th session, decreed by its decision number 64, dated 25/10/1398H, that treated and purified wastewater may be reused in wadoo' and drinking (Othman 1983).[1] That decree opened the way for the reuse of treated wastewater, which Saudi Arabia started to practice by using water from the Manfuha wastewater treatment plant in Riyadh to irrigate trees in the Dirab area west of the city. Other Islamic countries started to reuse wastewater in irrigation, at times untreated, a practice that triggered a variety of public health hazards.

Religious uncertainties have been removed by the above fatwa, and municipal wastewater has been targeted in many countries for treatment and reuse. Jordan has been a forerunner in this respect, with reuse confined to irrigation and power generation.

The Islamic tradition thus both supported the water tariff and collection of charges for servicing water and wastewater and endorsed the treatment and reuse of wastewater. These two aspects of water management are basic in the water policy of the country.

Access to Water and Wastewater Services

For a long time since the establishment of Jordan's Emirate of the Arab East in 1921, water supply to population centers has been the responsibility of municipal councils set up under the law. Likewise, electricity supply and other urban utilities, such as streets, curbs, parking lots, storm drainage, wastewater, and solid waste collection and disposal, were all part of municipal responsibilities. Water sources were usually in the proximity of population centers, until the population levels boosted the demand to the extent that it outstripped local resources and water had to be transferred from afar. Such transfers became familiar after the 1960s, when municipal water was transferred 40 to 70 km to Irbid and its villages from groundwater at Dhuleil. Smaller projects transferred water earlier, from Moses Springs in the Dead Sea catchment west of Madaba to the city in the 1930s. The transfers became regular starting in 1978, as outlined in Chapter 4, all to improve accessibility to municipal water. The policy regarding accessibility was always very clear: it was to supply water in pressure pipe networks to all residents within the town limits over which the municipal councils have jurisdiction. Central government stepped in to share with the councils the burdens of municipal water supplies (Chapter 2).

Building the networks to deliver water to the households was never a problem. All residents within town limits qualified for the connection to the municipal network, provided that the residence had been licensed for construction and approved for occupation. An occupancy permit is issued by the municipality to indicate conformity with the regulations and rules pertaining to construction. A water meter is installed to measure the delivered flow. In fact, the laying of water pipes in planned streets preceded all other municipal services in those streets in most cases. Rich and poor were accorded similar treatment, although influential persons may have gotten their connections installed faster.

Water delivery has been a function of resource availability, and the supply has been lagging behind demand since the 1970s. The pressing need for water prompted Amman to divert to municipal purposes the historic spring that sent fresh water flowing through the city, past the Roman amphitheater, and was being used for agricultural and environmental purposes. Orchards on both banks of the stream were thus denied their water, but fair compensation was paid to the owners for their water shares, and the lands of the orchards soon were included in the planning zones of the city, thus boosting their market prices.

The water networks supplied more than 97% of the kingdom's population by 1996. The 1980s had been pronounced the Water and Sanitation Decade in the wake of the 1977 UN water conference in Mar Del Plata, and that along with the institutional arrangements introduced in the kingdom and the attention the government paid to the water sector were instrumental in achieving this high rate of water network coverage. Network coverage did not guarantee continuous water service, however. Water shortages have made rationing a familiar policy measure since the mid-1970s.

The need for wastewater collection networks surfaced in the second half of the 1960s. Before the installation of wastewater collection networks, trucks emptied the full cesspools and hauled the solid and liquid wastes to designated disposal sites away from the cities. The cesspools did not function properly after the first filling, indicating saturation of the adjacent soils, and the need to empty them became more frequent. Additionally, the underlying aquifer, especially in Amman, received the untreated liquid wastes from the percolation pits. The concentration of nitrates in the aquifer, which was also used to supply the city with municipal water, became noticeable.

The policy then shifted to install wastewater collection networks and a wastewater treatment plant for the capital city of Amman. By 1968, the first wastewater treatment plant became operational at Ein Ghazal east of Amman. The wastewater collection network has been expanding ever since. Outbreaks of cholera in 1978 and again in 1981 prompted the government to focus attention on wastewater collection and treatment and on domestic water supplies. A sizable wastewater treatment plant at Khirbit As-Samra was hurriedly constructed as an interim arrangement before a permanent plant could be designed and built. But that interim plant became permanent and was overloaded within a short period of time because of an unexpected influx of people. The plant is being upgraded to improve its performance and the effluent quality.

Wastewater treatment plants spread to cover the major urban and rural areas in the kingdom. In addition to Amman and Zarqa, the populated areas of Abu Nseir, Baqa'a, Wadi Seer, Jerash, Irbid, Nuaymeh and environs, Ajloun and environs, Mafraq, Ramtha, Madaba, Karak, Tafileh, Maan, Wadi Musa and environs, Aqaba, and the rural middle Jordan Valley all have wastewater treatment plants in operation.

Social Impact of Accessibility to Water and Wastewater

The easier accessibility to water since the 1980s, especially in rural Jordan, had a profound social impact, particularly on women and young people. Hauling domestic water from its spring sources has been the job of women, assisted at times by their young children. That job alone consumed most of their time, especially if the water source was not a close-by spring. Children had to drop out of school to help their families with these and other chores, mostly related to farming. Even if the families owned animals that could be used to carry the water, the women had to haul it themselves because the animals were needed for farm labor. The easy access to domestic water enabled the women to perform other productive tasks and allowed families to afford the time and cost for their children to go to school.

The impact of accessibility to wastewater collection networks has been positive. The cost of emptying the cesspools and trucking the liquid waste to the designated sites has been growing. Alternative cesspools had to be dug to replace the ones that had become saturated with percolating liquids. The fees collected for connection to the network and the quarterly charges are less than the cost that would have been incurred otherwise. The impacts on public health and the environment have been positive. The treatment of municipal liquid waste is done at prescribed locations (the treatment plants) under the watchful eye of the officials of the Water Authority of Jordan (WAJ) and the Ministry of Environment.

An important issue concerning water access is the need for rationing. Because of the scarcity of water in the country, water is rationed in municipal and industrial networks and in government-operated irrigation systems. At the beginning of each summer, the WAJ announces a schedule of water distribution to the different zones of municipalities, when households will receive water for one or two days a week. This policy is unavoidable in light of demand. However, because of infrastructural and technical considerations, as well as social pressures and other difficulties, total equity in service is not attained. In most circumstances, influential neighborhoods, usually the wealthy, get better service than the neighborhoods whose voices can hardly be heard, usually the poor. To cope with the water shortage problems, families find themselves obliged to purchase drinking water from vendors who truck it in. The cost is high, at about US\$2.82/m^3, and it is very taxing to the lower-income families, who face more shortages than the upper-income families.

In the capital city of Amman, the poorer neighborhoods in the east, south, and north receive good-quality water drawn from aquifers, with the only treatment being disinfection. The western parts of the city, with more well-off residents, receive treated surface water from the Jordan Valley. The quality of the former is better than that of the latter.

Rationing is also used as a management tool in irrigated farming in the Jordan Valley to confront the water shortage. Measures associated with shortages give priority of service to perennial crops—citrus fruits and other orchards—to protect the relatively high capital invested in them. Seasonal crops, usually the property of tenants and poor farmers, receive rations commensurate with availability. Social equity is sought so that poor farmers will get a portion of the water to sustain their crops while the capital-intensive crops are served with enough water to avoid irreversible damage. Fortunately for water managers, few seasonal crops are grown during the season of high water demand. Only at the beginning of the growing season, from early September until mid-October, does the water shortage have a perceptible impact on seasonal crops, whose demand for water in the Jordan Valley increases during the winter months, when water is more available. Some rationing also may be necessary during May and early June.

Another important issue is accessibility by people in the squatter areas and other unlawfully built premises to municipal water. In order for a household to have utility connections, a permit to occupy the premises must be issued by the municipality. This occupancy permit is issued only if the dwelling has been built in accordance with the building code under which a building permit was issued in the first place. Any violation of the code is an obstacle to receiving an occupancy permit, without which the premises may not legally be connected to the various utility networks—water, electricity, wastewater, telecommunications, and so on. In spite of this regulation, persons whose premises are denied legal connection to the water service often work out an arrangement with a neighbor who is legally connected and pipe water to the residence unlawfully. With the block water tariff, the water thus metered is the consumption of more than one subscriber, and the tariff goes up to the disadvantage of the legally connected subscriber. The end result is a higher cost of piped water that these poor people have to cope with.

The case of Palestinian refugee camps is different. No occupancy permits are required for these temporary shelters put up in times of emergency, and they are connected to the water network. Wastewater networks also have been built for the crowded refugee camps of Baqa'a, Schneller, Hussein, and Wehdat. The rules that are enforced in urban areas are not applied to refugee camps.

Equity Issues in Water Allocation

Issues have arisen over equity in water allocation among the different uses: agricultural, municipal, industrial, and environmental. The government has attempted to mitigate these inequities through rationing of service, tariffs, and subsidies.

Agricultural versus Municipal Uses

Agricultural uses of water constitute the majority share in water resource allocation. Although this is justified on the premises of society's need for food production and other social, environmental, and demographic benefits, urban dwellers look with envy to agricultural users of groundwater, especially when the aquifers are close to the urban areas. Urban dwellers do not see equity when they are facing shortages in municipal supplies while cucumbers and tomatoes are being irrigated with precious groundwater. Agricultural water users counterargue that they have developed a lifestyle around irrigated agriculture, and their lives would be disrupted if water is taken away from them. Compensation for annulling the water extraction permit, legally issued in the 1960s when water stress was not that high, does not solve their problem of dislocation and the corresponding need to build a new lifestyle and take up other jobs they are not trained for.

It can be rightfully argued that agriculture, a tradition deep in Jordanian and Middle Eastern history, provides a cheaper capital cost per job created than does industry and some services. However, it consumes the maximum water flow per job. In social, environmental, and political terms, agriculture is strategically important not only for Jordan, but for countries of the world at large. Agriculture allows a balanced population distribution over the territories of the country, provides jobs in direct agricultural engagement and a score of supporting services, and enhances the environment.

A definite advantage that agriculture retains over most other sectors is the possibility of its use of treated wastewater. Although the primary gains of wastewater collection and treatment are related to public health and the environment, the land application of wastewater is a plus in water-short countries like Jordan. Farmers, long used to irrigating with fresh water in the Jordan Valley, complain bitterly as they see it diverted away from their farms to municipal uses in Amman and replaced with treated wastewater. They argue that water is a generator of wealth, and Amman is wealthier than the Jordan Valley. Thus the diversion of water resources from the valley is in effect a measure to redistribute wealth in the country in favor of the rich. They cite as evidence the fact that the consumers of the Jordan Valley water in Amman live in western Amman, the wealthiest neighborhood in the country. "What is this policy of water allocation," they ask, "that takes water away from rural producers and allocates it to people in urban areas for taking showers at a time when the policymakers claim they aim at halting the migration from rural to urban areas?"

"People," they rightly reinforce their argument, "follow the water. There is no life without water. But instead of people coming from Amman to live in the Jordan Valley, where the water is, government is pumping the water from the valley up the steep mountains to Amman and returning wastewater for the use of rural areas instead."

Industrial Water

Industry has priority in water allocation. A licensed factory will be issued a license to drill a well and abstract groundwater, even if the aquifer is overused already. Permits for agricultural uses have come to an end; no such permission is given to drill wells for agriculture. The justifications for industry's priority are the number of jobs it creates per unit flow of water, its role in import substitution, the desire to attract foreign investment, and the enhancement of export to earn foreign exchange. A similar priority is given to educational institutions to support the government's policy in promoting and encouraging investment in education.

Environmental Water

Environmentalists are complaining that diversions away from the natural watercourses and overabstraction of groundwater have had damaging effects on the environment. They rightfully argue that the environment has not been paid the attention it deserves among the different water-using sectors in the country. Groundwater quality degradation, disappearance of certain aquatic species that once existed in surface water, the shrinking of desert oases because of overpumping, water pollution, and other deleterious effects on the environment brought about by users of water resources have been clear environmental losses. In fact, environmental awareness in Jordan was not emphasized before 1980, when care for the environment was entrusted to the Ministry of Municipal and Rural Affairs. In 2003, it was accorded a separate cabinet portfolio.

The involvement of environmentalists and environmental nongovernmental organizations (NGOs) has changed the outlook for environmental water. A major water project for the development of the Southern Ghors through irrigation, the Mujib–Zara–Zarqa Ma'een–Hisban Project (described in Chapter 4) was modified because of pressures from environmentalists and NGOs. The environment was allocated a share of the Mujib water to sustain the aquatic life in that stream, at the expense of the cost of use of that water, especially in agriculture. In the same project, a recharge dam was built across Wadi Wala, a tributary to the Mujib, to feed the aquifer that had been tapped to supply Amman with municipal water. The supply to Amman dried up precious springs fed by the aquifer, with all the attendant environmental consequences. The recharge dam, Wala Dam, resurrected those springs in 2004.

Mitigation of Inequities

Some inequity issues concerning water allocation are offset by water tariffs. Agricultural water tariffs are subsidized in the Jordan Valley but not outside the valley, except for modest subsidies accorded for electrical energy to pump agricultural water. Well owners in the Highlands carry the costs of operation, maintenance, and replacement of their systems and are now charged for any quantity pumped in excess of 150,000 m^3/y. Industrial water tariffs are high, and so are the charges for hotel and tourism uses.

Agriculture, which occupies third place in water allocation, has been favored in subsidy policy. In addition to the subsidy for irrigation water in the Jordan Valley, agricultural income is exempted from taxes, and energy consumption for irrigation is subsidized as well. Thus the lower priority is mitigated by the subsidies agricultural production currently receives.

Policy on Wastewater Treatment and Reuse

Use of untreated wastewater is prohibited by tradition and municipal regulations. For a time, treated wastewater was used to produce plants eaten raw employing surface irrigation methods, but when cholera broke out in 1978, the prime minister resorted to his power as general military governor of the country to ban those practices. Fields that were using wastewater and surface irrigation methods were plowed by a national defense order.[2]

Treated wastewater, blended with freshwater resources, was later approved for use in irrigation employing drip irrigation methods and advanced farming technology. This was approved in the Jordan Valley where drip systems were used and plastic mulch and plastic tunnels and houses were used in agricultural production.

The pressing need for municipal water, especially in Amman and environs, and the elevation that city has relative to the Jordan Valley, prompted planners of valley development to introduce the idea of using water more than once.[3] Water could be pumped up from the Jordan Valley to cities in the Highlands, used in domestic purposes, collected and treated to acceptable standards, and left to flow back to the Jordan Valley by gravity for reuse in irrigation. The King Talal Dam, on the Zarqa River, would be the primary regulator and would allow the blending of treated effluent with storm water impounded by the dam. The idea was debated in a high-level meeting chaired by the king (Chapter 4) and was approved in 1978. The importance of the issue was reflected in the high-level representation in the meeting that debated the policy.[4] Irrigation water was thus augmented with treated wastewater, which blended with fresher water in the King Talal Dam reservoir. It became possible both to sustain irrigated agriculture and to allocate agricultural fresh water to municipal uses after the Deir Alla–Amman project was completed in 1986. Downstream farms in the valley would receive fresh water during the wet months from the Yarmouk River, whose flow was not regulated, and during the balance of the year, a blend of treated wastewater and storm water.

Equity issues, as outlined above, were raised by farmers of the Jordan Valley and are still debated by the development planners. In addition to the arguments they presented, the debate questions the fairness to downstream farmers, who receive mostly treated effluent to meet their irrigation needs in the dry months. The question becomes pertinent especially in drought years, when the flow of freshwater resources feeding the irrigation system in the wet months diminishes. The quality of the effluent returning to the valley is inferior to the quality of the fresh water that is pumped from it for domestic use in Amman. The salinity and alkalinity are higher, adversely affecting the productivity of the water and limiting the range of crops that can be planted using it. This has negative impacts on the income that the farmers get for their annual work and investment, the net worth of the farm products in the long run, and the market value of the agricultural land. Planners argue, conversely, that the effluent contains nutrients that enhance the productivity of soils and the yields of crops. Nevertheless, the farmers argue, it is not fair to charge them the same tariff for their irrigation water as farmers upstream who use freshwater resources.

The social and health concerns on the part of regional countries that import the valley's produce resulted in a ban on import of Jordan Valley vegetables for a number of years. The measure was not without political underpinnings resulting from the invasion of Kuwait by Iraq and the ensuing Jordanian stand.[5] This measure unfairly impacted all the farmers of the valley, both those who were served treated effluent (20%) and those who were not (80%). Bans were relaxed after site visits confirmed the adequate treatment of effluent. Guidelines of the World Health Organization and the Food and Agriculture Organization were observed in the process of reuse. Citrus fruits have been allowed into the markets of Saudi Arabia, but not vegetables. In October 2004, the Saudis dispatched a team to look into the possibility of lifting the ban on imports from the Jordan Valley (*Al Rai* 2004).

The economic, social, cultural, and political gains from water resource development can be enumerated and detailed theoretically, but a better way of presenting them is to review a real-life development story that had water at its center, with integrated social and economic development as its goal. Such is the case of the integrated development of the Jordan Valley.

Water for Economic and Social Development

A major shift in government policy on water allocation and use happened in 1957 in the midst of political and security turmoil in the Middle East. Jordan had negotiated termination of the bilateral treaty with Great Britain in 1956, dispensed with the services of British officers in its armed forces, and looked toward Arab neighbors for military coordination and financial support. The outcome was not very encouraging.

Jordan then approached the United States for financial assistance to buttress its efforts in economic and social development. The country had received the major portion of Palestinian refugees and united with the West Bank, adding more refugees to its population. It was badly in need of development, job creation, and economic growth. Water for irrigated agriculture in the Jordan Valley was the answer. The United States had come up with the Unified Plan for the development of the valley in 1955 (Chapter 10), and Jordan's share of water under that plan was sufficient to irrigate 36,000 and 16,000 ha east and west of the Jordan River, respectively.

The East Ghor Canal Project

The United States came to Jordan's aid to implement the irrigation scheme in stages. It hoped that the irrigation project would create enough opportunities for settlement of Palestinian refugees in the valley until their problem was finally resolved. Irrigated agriculture was suited to the cultures of Jordanians and Palestinians and the resources at hand.

Both the donor and the recipient country, Jordan, wanted social equity to be observed in the process of development. A master plan was formulated for the development of the Jordan Valley using the water flow of the Yarmouk River, the major tributary to the Jordan, in an amount not exceeding Jordan's share in the Unified Plan. Jordan's policy on the whole development process was elaborated in legislation that, for the first time in the history of Jordanian society, introduced land reform. In essence, the legislation made the government a partner with the landowners, especially for the lands that could not become productive without government investment in irrigation systems. Under the legislation, owners of 5 ha of land or less would retain their ownership in irrigated land after the irrigation project was implemented. Those who owned more than 5 ha would get 5 ha plus 25% of the area in excess of 5; those who owned more than 10 ha would receive 6.25 ha plus 17% of the area in excess of 10 ha, and so on, until the owner of 100 ha or more got to retain only 20 ha. The rest of the ownership would go to the East Ghor Canal Authority, a government agency that was set up especially to implement and manage the project. The land that the authority got plus any other land registered in the name of the Treasury would then be distributed to landless farmers and professional farmers owning small land holdings. The transactions of any sale on irrigated land in the valley were to be approved by the authority before they were validated.

Those landowners who were growing perennial crops using water from side wadis were able to retain the total area of the land they had developed with these crops, but the redistribution provisions of the law still pertained to any land they had been using for seasonal crops. This applied to only a limited number of landowners, however, because the water they could use for perennial irrigation before government intervention was limited. For landowners to relinquish part of their ownership in return for perennial irrigation of the land they retained was to them a good deal. The part the landowner retained would be worth a lot more than the entire piece of land that had produced very little, if at all.[6] In effect, the law maintained a balance in attention paid to landowners and landless farmers through government investment.

The East Ghor Canal Authority managed to put under perennial irrigation a total of 11,400 ha between 1959 and 1966. Its successor, the Natural Resources Authority, increased that area by 800 ha by 1969. No further expansion was possible after this because of the aftermath of the 1967 war, which caused instability in the Jordan Valley and the country at large until 1972.

Integrated Social and Economic Development

The vision for the valley's development with water and irrigated agriculture as its backbone took on a more comprehensive outlook in 1972, after law and order were reestablished in the country. Legislation was enacted for that purpose, and a new organization, the Jordan Valley Commission (JVC), was created in 1973 for the task. In 1977, another legal entity was established, the Jordan Valley Authority (JVA), which became the successor of the JVC, the Jordan River Tributaries Regional Corporation, and the departments in the Natural Resources Authority and the Domestic Water Supply Corporation in charge of the Jordan Valley services.[7]

Keeping the provisions of land redistribution, the government policy shifted to pursue, through water resource development, an integrated social and economic development approach. The Jordan Valley Development law number 18 for the year 1977 prohibited the sale of developed agricultural land to any party other than the JVA. Land reform expanded to include the redistribution of residential lands within planned towns and villages. The responsibility of government in this new partnership was the provision of such things as utilities, housing, schools, health centers, streets, and parking lots.

Economic and Social Impacts of Water Resource Development

The economic and social impacts of integrated development, over a span of a dozen years, are reviewed in this section, based on a thorough study financed by the United States Agency for International Development (USAID) in 1986 and 1987 (Shepley et al. 1987). No similar studies were conducted thereafter, denying planners and decisionmakers the opportunity to assess the impacts of policy changes and of diluted emphasis and little attention. In a recent countrywide sample study of poverty, certain areas of the Jordan Valley, Deir Alla, and Safi, were cited as having the highest rates of absolute poverty (DOS 2005). Other press reports have carried news of poverty and need in the northern Jordan Valley, the portion with the longest tradition in its development.

The planning process for the Jordan Valley began in an effort to realize its vast agricultural potential, as it is considered a natural hothouse capable of producing winter fruits and vegetables. Government investment in the development of the valley added up to US$24.7 million by 1969 and US$260 million by 1982. It increased to US$660 million by 1987, resulting in a dynamic transformation in the economic, social, cultural, and political conditions of the Jordan Valley. A total of US$260 million was invested in water resource development in the form of dams and irrigation networks, resulting in the irrigation of 26,800 ha. The remaining US$400 million was invested in village development, including the construction of all needed social infrastructure, such as schools, health centers, administration buildings, streets, housing, highways, and marketing centers.

Reversing the Rural–Urban Migration Trends. The transformation of the Jordan Valley agricultural sector from subsistence to a dynamic market orientation, with corresponding rises in productivity and income, has caused it to become a social magnet. Since 1973, there has been a net immigration to the valley, thus reversing the familiar trend of emigration from rural to urban areas. The population of the valley increased from about 63,000 in 1973 to 125,000 in 1986 and 240,000 in 2000.

The reasons for the net immigration seem very clear. From 1978 to 1986, the weighted mean per capita income increased from US$381 to US$601. The per capita agricultural GDP of the valley in 1979 was US$415, compared with the national average of US$265. In 1986, the per capita agricultural GDP in the valley rose to US$713, compared with the national average of US$586. Thus the rural incomes in the valley were consistently higher than the national average, and certainly higher than in other rural areas in the kingdom, a fact that stimulated immigration to the Jordan Valley.

Economic Growth and Structural Changes. Improved income trends have lowered the percentage of the valley population below the poverty line (assumed to be US$212/cap/y in 1986) from 22% to 20% between 1978 and 1986. The immigration of people to the valley in search of economic opportunities has created a corollary demand for goods and services to support the population influx. This growth in demand produced structural changes in the regional economy. From the early 1970s to the late 1980s, the economy of the Jordan Valley has progressed from a primarily single-sector economy (agriculture) to a more diversified multisectoral economic structure, wherein output of nonagricultural services and some goods has accounted for an increasing share of the regional GDP. In 1978–1979, agriculture supplied 51% of the regional GDP, compared with 5% of the national GDP. By 1986, agriculture's share of the regional GDP had fallen to 35%, compared with the 7.2% of the national GDP.

The Social and Cultural Dimension of Water Policy. Along with the structural changes in the regional economy of the Jordan Valley, significant parallel changes also occurred in the regional employment distribution during the period 1978 to 1986 and thereafter. In 1978–1979, males constituted 72% of the Jordan Valley workforce, which totaled 27,126 persons; 81% of these males (15,820) were employed in the agricultural sector, forming 73.5% of its workforce, including the majority of hired farm workers in the valley. At the same time, female farm workers numbered 5,690 and made up 26.5% of the agricultural workforce. Of the female workers, 96% were unpaid family agricultural labor.

By 1986, the total Jordan Valley workforce had increased to 58,930. The total agricultural labor force was now 25,340, of which about 22,000 were expatriate

Egyptian labor; the remaining 3,340 were supplied from the Jordanian female farm population for paid seasonal harvesting and weeding. By this time, the incidence of unpaid family farm labor was virtually nonexistent. The percentage of the male workforce in agriculture had declined to 43%, down from 58.3%. For females, the decline was even more dramatic, from 21% to 5.6%. Three significant trends were seen in male and female employment between 1978 and 1986: the virtual disappearance of native Jordanian males from the hired agricultural labor force by 1986, replaced by migrant workers from Egypt and Pakistan; the shift in female employment from predominantly unpaid family agricultural labor in 1979 to paid employment in education, health, commercial, and service professions outside the home by 1986; and the growth in demand for increased wages among females still working in the agricultural sector by 1986. Between 1978 and 1986, the incidence of nonagricultural employment in the Jordan Valley increased by an aggregate 100%, with public employment increasing by 163%. During this period, the number of private commercial businesses in the valley increased from 489 establishments to 938, an increase of 92% in eight years.

Social Infrastructure and Services. A key component of the Jordan Valley's integrated development strategy was the inclusion of provisions for social infrastructure development: education, health, public utilities, housing, communications, transport, and town planning. Both public and private sector institutions have played roles in the development of the valley's social services.

The infrastructure and social services installed in the valley from 1973 to 1986 have provided levels of service that have been at least equal to, and in many cases exceeding, those in the rest of the country. For example, the student–teacher ratio in the valley in 1986 was 1:21; the national average was 1:25. The number of students per classroom in the valley was 26.5, compared with the national average of 33.3. The valley's health facilities were also superior. Jordan had 2.16 clinics per 10,000 people, whereas the valley had 2.92. The national average was 1.81 pharmacies per 10,000 people; the valley had 3.39. The number of doctors per 10,000 people was 7.52 for Jordan, and for the valley, 8.89.

There were many other illustrative indicators of the impact of social service development. The housing spatial density increased from 6.2 m² per person in 1973 to 14 m² per person in 1986. In 1973, 10% of the residential dwellings in the Jordan Valley had access to potable water and electricity; by 1986, it was 100%. The number of telephones per 10,000 people increased from very few to 66.5, compared with the national average of 65.8.

The introduction of clean water networks to the villages and towns in the valley reduced the rates of child morbidity (the incidence of disease) and child mortality. It also saved the time that females of the family had spent to fetch their daily needs of fresh water from springs or tube wells. More often than not, these chores had kept girls away from school. Now the enrollment of girls in schools increased drastically.

The valley was connected to the national electricity grid, and because of their increased incomes, families in the valley were able to obtain electrical appliances, including television sets. These helped spread education among the public and in some cases helped increase the duration between pregnancies. The valley was further connected to the national telecommunications networks. The impact this service had on society was immense. It became easy to call for medical help or for civil defense services, as well as the capital, where agricultural marketing was done. Farmers did not need to travel to Salt or Amman to get in contact with people.

Social Dynamics. An analysis of social dynamics provides a true measure of developmental impact on changes in the quality of life. In the Jordan Valley, indicators of social change show that society is becoming less tradition bound and more outward-looking, and it is placing educational achievement and material progress above more traditional tribal values of family ties and conventional male–female roles in the socioeconomic order.

In the early 1960s, most marriages in the Jordan Valley were patrilineal parallel cousin marriages, the traditional marriage arrangement within the patriarchal tribal structure found historically on both the East and West Banks of the Jordan River. The primary roles of women in society were to perform domestic chores of the household and produce male heirs. A random survey in 1986 of 40 households throughout the Jordan Valley showed that a significant shift in outlook toward the traditional marriage structure was in progress. More than a third of the marriages in the sample were exogenous, outside the tribal lineage. The role of women was shifting noticeably from domestic servant and childbearer to a more integrated and fuller participant within the socioeconomic structure. This phenomenon has been reinforced even more noticeably since the beginning of the 1990s because of the sudden devaluation of the local currency and the subsequent jump in the cost of living. Women, daughters and wives alike, have been assuming more of an economic role to help provide for the family budget.

In 1986, more valley females were attending school than males, constituting 61.3% of the student body. University education was desired by 55% of parents for their sons, and 57% for their daughters. In a single generational span, the average number of years of education for women increased 130%, compared with the male average of 91%. Female literacy during the same generational span increased from 48% to 95%, and for males, from 70% to 100%.

Jordan Valley women are spending more years in school, greater numbers are leaving the home for paid professional employment than ever before, and the mean age of marriage in the valley has increased from 17 to 19 years. This trend shows an increasing tendency of Jordan Valley women to delay marriage to further educational and career goals. The trend has been evident in recent years by the number of females from the valley who have enrolled in universities and institutions of higher learning to gain a better edge in the job market.

Commensurate with changes in social structures and educational effectiveness, a sample survey of 40 households throughout the valley conducted in 1987 manifested other positive indicators of life quality changes that reflect favorably on the impact of improving health care and rising incomes in the Jordan Valley, all made possible with water availability. In 1961, the life expectancy index for Jordan was 45.8 years for males and 46.5 for females. By 1987, the national index had risen to 67 for males and 71 for females. In the Jordan Valley, the index was 5% higher than the national average, at 70 years for males and 75 for females.

Other health quality indicators offer equally convincing evidence of the effectiveness of health services in the Jordan Valley. In 1961, the infant mortality rate for Jordan was 151 per 1,000 live births. In 1986, the national average for Jordan, including the major urban areas of Amman and Irbid, was 60.4 per 1,000 live births. In the valley, which is predominantly rural, the 1987 rate was 55.

The morbidity incidence for the Jordan Valley in 1987 was 17 per 1,000, versus a national average of 18 per 1,000. The combined effects of declining morbidity and infant mortality rates and increasing life expectancy pushed the average family size from 5.7 in 1973 to 7.5 in 1987. As a result, the structure of the demographic profile was shifting toward the lower end of the age spectrum. In 1973, 53.2% of the population was under age 15. By 1987, the percentage had climbed to 59%.

Political Development

The Jordan Valley falls within several electoral districts in the kingdom. In 1989, the northern valley and a neighboring administrative district formed one electoral district to elect two seats in Parliament; the middle Jordan Valley was part of a wider electoral district, the Balqa', which was represented by seven members of Parliament; and the Southern Ghors (south of the Dead Sea) was part of the Karak electoral district, whose share was nine seats in Parliament.

Prior to the integrated development of the Jordan Valley, its voters could barely vote in one deputy to the lower house of Parliament, usually from the Middle Ghors (South Shuna). They always formed alliances in favor of politicians who ran for election from outside the valley. The situation transformed after the valley's integrated development. In 1989, for the 13th Parliament, three members were elected from the Jordan Valley: one from the northern valley and two from the middle valley. In the 1993 elections, the number of valley deputies rose to six, including two from the Southern Ghors. In the same period, the cabinet had two ministers from the Jordan Valley, and coupled with the parliamentarians, that was a quantum political leap for the valley indeed.

Today the valley has asserted its political character. It has three separate electoral districts, covering the northern, middle, and southern Jordan Valley. Four deputies occupy seats in the lower house of Parliament, and two were appointed by the king in the upper house. Participation of valley residents in the cabinet has become a familiar event in the political life of the country.

Attitudes, Expectations, and Case Studies

One of the heads of household in a 1987 sample survey summed up his condition by saying, "God's bounty has been bestowed upon me and my family, what else could I wish for!" He added, "I succeeded in farming; my sons are all doctors in Amman; my granddaughter is a headmistress of a school; my wife has started literacy classes; and I am at peace with my neighbors and community."

Another example of fulfilled aspirations can be seen in the case of a 57-year-old middle valley woman, who was proud because she had just finished the elementary cycle by attending literacy classes. She wanted to continue until she obtained a Tawjihiyya (high school diploma). In her own words, she had realized her "life long ambition of being able to read like [her] husband and children and of being aware of what was happening in the world."

A young valley woman who made her living from sewing wanted to run for Parliament. She said, "Come back five years from today and I will be sitting in the Parliament House in Amman, not at this sewing machine."

A young male employee of tribal origin was studying for his master's degree in administration at the Jordan University. He had to commute three times a week by bus to Amman. He said, "My younger brother is at Alexandria University, Egypt, studying to become a lawyer, and my parents want me, their eldest son, to do better."

Of the sample surveyed, 50% would allow their sons to migrate in search of education and better jobs, although 90% of the homeowners were unable to live elsewhere and wanted to die in their place of birth.

The above are just a few of the many examples of how people living in the Jordan Valley expressed their optimism about their lives, opportunities, and ambitions, and the resounding endorsement they gave to the quality of life they had come to experience. They fully appreciated the social dynamics that had taken place in the valley over a decade and a half.

Supporting Capacity of Water in the Jordan Valley

In 2000, the Jordan Valley, with a population of 240,000, used 200 mcm of irrigation water. This means that a unit flow of water, at 1 mcm/y, was capable of supporting 1,200 persons. Given a family size of 7.5 persons, a unit flow of water can allow 160 families to settle in rural areas.

It was shown in 1987 that a farm unit of 3 ha was capable of supporting two families working in direct agricultural jobs. A unit flow is capable of irrigating an average of 160 ha of horticultural products, or about 50 farm units. These would support 100 families whose heads work in direct full-time agricultural labor. There will be 60 more families working in supporting services and in part-time agricultural labor.

This supporting capacity of water is a very important factor in the country's population distribution. It means that for every unit flow of water diverted from the rural Jordan Valley to urban areas with nothing returned for reuse, 160 families will be deprived of their means of livelihood and have to follow that water to urban areas. In case of the already implemented scheme of water supply to Amman, about 60% of that water returns to the Jordan Valley, and thus the operation denied only 40% of the number of families the 45 unit flows of transferred water would have allowed, or 2,880 families. Those families instead had to find opportunities in Amman, to which the water was transferred, and cost the government in social infrastructure and services, employment, or unemployment.

Subsequent Changes in the Markets

The sentiments expressed by people in the sample survey of 1987 may not be the same today (2006). Marketing difficulties have clouded the bright achievements of the Jordan Valley development. Additionally, political developments in the region intersected with trade and disrupted the smooth flow of goods and services between Jordan and the Gulf states. The purchasing power of consumers in the region has diminished because of the cost of wars, declining oil prices over the period 1985–1995, and the need for higher security spending. Jordan Valley's agricultural products were particularly hit, and revenues declined drastically.

The net income from a unit flow of water has decreased, and the need for improved marketing mechanisms is pressing. Until then, agricultural contribution to the national GDP will continue to be low, but the chain of dependencies on agriculture, and irrigated agriculture at that, continues to be substantial.

Water in Industrial Development

The above analysis of the Jordan Valley development case is not meant to imply that water resource development in irrigated agriculture has the maximum economic and social returns. Undoubtedly, water development for industrial purposes has its own returns. Jobs are created in industrial development with much less water consumption than is needed in irrigated agriculture. Whereas 1,000 m³/y per job has been an average in the Jordan Dead Sea industries, the water needed per direct job in irrigated agriculture has been about 12,500 m³/y. When jobs in supporting services are added, the water per agricultural job drops to 8,300 m³/y. Conversely, the capital investment per job in industrial development in Jordan was about US$26,000 in 1986, compared with a direct investment of US$7,500 (in current prices) per job in irrigated agriculture, including the construction of dams.[8] In contrast to the economic output of agricultural water, the output from a unit flow of industrial water is much higher. Other input parameters are not comparable either. The social benefits from agriculture and the corollary benefits gained from it in dependent sectors make water development in agriculture attractive to society, especially if the traditional activities of the rural areas have been in agriculture. It creates a culture of its own that is distinct from industrialized urban communities.

Equity Issues in Water Tariffs

Cost recovery of water services became a central issue in the economic restructuring program that the government embarked upon in 1990. The debate that ensued covered municipal, industrial, and irrigation water. Addressing the cost recovery was one criterion by which donors assessed the feasibility of making new loans to water projects.

The government's policy was set to recover the operating and maintenance costs of water and as much as possible of the capital cost of water projects. In 1997, the Council of Ministers issued a municipal water tariff. A low tariff was set for low consumption rates, to be subsidized by the higher consumers. Any consumption rate above 20 m³ per connection per quarter, determined to be the average rate consumed by the poor, is not subsidized. If the total consumption per cycle falls within the consumption rate of the poor, then all of it is subsidized, but if it is higher, no fraction of it receives any subsidy. A ceiling charge of US$1.41/m³ is set for high consumers, and this rate applies to tourism uses in hotels and other facilities. About 55% of the water charge is added to pay for wastewater collection and treatment services. This tariff structure establishes a cushion for the poor and manages demand via escalating charges for higher consumption rates until the ceiling is reached.

Regarding irrigation tariffs, the policy has been to recover the operating and maintenance costs of the service. A higher tariff of US$0.05/m³ is set for monthly consumption rates of more than 4,500 m³, and a rate of US$0.028/m³ is charged for a bracket of consumption falling between 3,500 and 4,500 m³. Higher consumption rates are applicable to perennial crops, usually planted by more well-off farmers. The tariff, in short, favors the seasonal crop growers, and these are usually the less well-to-do farmers. Additionally, the orchards require more expensive water, stored behind dams, for irrigation during the dry months, unlike seasonal horticultural crops, whose demand is mostly met by allocation from unregulated flows of the Yarmouk.

The structure of water tariffs is thus commensurate with the requirements of equity: the more well-off farmers, who usually consume higher flows of expensive water to irrigate orchards, pay higher charges, and the less well-to-do farmers who use less water pay lower charges.

The Social Impacts of Water Management Adjustment Programs

The government adopted adjustment programs for the management of municipal and irrigation water service in the Jordan Valley. It initiated this policy in 1999, entrusting the operation and maintenance of water and wastewater service in the Greater Amman area to a private company. The water and wastewater services in the Aqaba region have been transformed to a publicly owned company as a step to make it a private company. Tenders have been floated to award management contracts for the water and wastewater services in the northern governorates, and the programs are still ongoing.

For about three decades prior to the water adjustment policy, municipal water shortages dictated that service of piped water had to be rationed, and a rotation system of service was adopted. Each neighborhood in Amman, for example, would receive water once a week. The day of water service became a day of alert in the household. Laundry, cleaning, irrigating the plants, taking baths, and storing water in household tanks had to be scheduled to take place on that day. Women were the "generals" in that battle, and they had to postpone all other duties to manage the water-day events. Poor households did not have the capability to install water storage facilities and mostly had to rely on the water supply of one day, storing as much as they could in buckets and clay jars for drinking purposes.

If the water supply fell short of meeting the needs for one week, or for the span of time before the next water service in the rotation system, families usually relied on buying water from vendors at a much higher price than the poor or middle-income families could afford to pay.

The involvement of the private sector in the management of water and wastewater service in Greater Amman improved the quality of service and provided water to households by trucks if the cycle of service did not supply sufficient water or was delayed. The new management system was concurrent with public investments by WAJ to replace old water networks and upgrade the wastewater collection and treatment system. The combined impact of these two undertakings supplied relief to the women of the households and saved time that they could spend on other productive responsibilities. The financial burdens associated with purchasing water from vendors were alleviated under the new system of water service management.

Adjustment programs were also implemented in agriculture and had an impact on farmers and farm production. A study was conducted jointly by Jordanian and German experts to evaluate such impacts on farmers, including their perceptions of and reactions to the adjustment programs (CATAD 1996). The study addressed rain-fed, irrigated, and livestock farming systems. The findings are summarized below.

Rain-Fed Farming Systems in the Highlands

Until the 1960s, the predominant way of farming in Jordan had been rain-fed agriculture, on which the livelihood of many families depended. Family labor worked the land of various sizes of holdings. As farm income did not increase along with the cost of living, the younger generation attempted to migrate to urban areas in search of jobs in industry and services. Then, in the early 1960s, the government started to purchase the produce from rain-fed areas, especially wheat and barley, at subsidized prices. It had dual objectives: to support rural areas and maximize the production of strategic crops.

The agricultural structural adjustment programs designed in the early 1990s called for the elimination of government subsidies. This policy measure negatively impacted farm income and reduced the ability of farming families to meet the cost of living with agriculture revenue. The gap between the cost of living and family income resulted in a decrease in years of schooling, especially for girls, and opened the way for poverty to creep in. Many farmers in rain-fed areas found themselves facing living obligations that outstripped their agricultural income and had to sell the land from which they and their forefathers had made their living.

Irrigated Farming Systems

The water stress triggered a trend among farmers who engaged in irrigated agriculture to adopt advanced irrigation systems to make the maximum use of the supplied water. This measure entailed capital investment to purchase and install drip irrigation networks, as well as additional operating costs incurred by pumping the water in the farm networks. Increased agricultural yields helped justify the added investments.

Irrigated farm incomes were attractive until the mid-1980s, when declining oil prices reduced the purchasing power of export markets in the Gulf, and Jordanian agricultural production saw increased competition from products shipped from Lebanon, Turkey, Syria, and Greece. Shrinking markets as a result of the Gulf War exacerbated the plight of Jordanian farmers. In 1981, the government began to intervene in the market of fresh tomatoes, purchasing the excess produce for industrial purposes at subsidized prices.

The adjustment programs called for an increase in the water tariffs and the elimination of government intervention in the production and marketing processes. They put a ceiling on the abstraction of groundwater, especially in the Highlands, to reach an equilibrium state between recharge and abstraction, and they required the gradual reduction of abstraction from fossil nonrenewable aquifers used in agriculture. The combined effects of increased production costs, shrinking markets, limitations on abstractions, fines associated with violations, and the lifting of government subsidies negatively impacted farm incomes. The adjustment programs also called for the deregulation of the land market in the Jordan Valley, and free sales have been permitted by law since the year 2000.

The depressed farm income in comparison with the increased cost of living started to create distress among the populations of irrigated agriculture areas. Poverty started to grow, and the momentum of social and economic gains triggered by integrated development slowed down. Another result has been an increase of farm land sales, mostly in the Jordan Valley, from poor farmers to wealthy people, a process that would encourage rural–urban migration, with the associated burdens on society.

Livestock Farming

Structural adjustment also has affected livestock farmers, especially in the low-rainfall zone of the Badia. This area receives, on average, between 50 and 200 mm of rainfall per year. The Bedouin population that lives here is primarily engaged in raising livestock, and their life is characterized by seasonal migrations of the entire family to the best rangelands. The nomadic system is interrelated with rain-fed areas in the Highlands and irrigated agriculture in the Jordan Rift Valley. The nomadic pastorals move their flocks in winter to graze on early ephemerals in the desert rangelands. In late winter and early spring, they take their flocks to lower lands. Flocks are then moved in late spring to early summer to agriculture cropping areas, where they can graze on crop residues.

The government has a standing Drought Committee that helps livestock farmers cope with stress in years of drought. It supplies animal feed at subsidized prices, provides free water for animals, and at times, regulates the market of live animals.

The requirements of the adjustment programs aim at eliminating subsidies, of which support of the livestock farmers is part. Livestock in the Badia constitutes the majority of economic activities that the less fortunate residents there conduct. Their incomes will be suppressed by the increased cost of production they have to face. Livestock growers in the Highlands and the Jordan Valley often take up other economic activities to supplement their incomes, and the green fodder in the grazing areas helps provide for the animal feed. This advantage is not always available in the Badia.

When rainfall prospects diminish, livestock farmers, especially in the Badia, have been dumping their livestock in the market because of the increase in the cost of production. Low rainfall reduces the fodder in the natural grazing lands, and farmers have to purchase animal feed at unsupported prices. This has had profound impacts on the standard of living of livestock-farming families.

Summary

The people of Jordan are of several ethnic origins but are part of the Islamic culture. Their religions are Islam and, to a much smaller extent, Christianity. Water has a special status in each, thus assuming a role in cultural heritage. In Islam, people are partners in water; it cannot be sold unless an effort has been made to provide it. A fatwa was issued permitting the reuse of treated wastewater.

Central government involvement became pronounced when local resources could not cope with municipal demands. Water transfers were beyond the capacity of municipal councils, and government stepped in to build municipal water transfer projects. Such transfers appeased urban areas at the expense of rural areas, where the transferred water could have been used in irrigation. Water networks supplied 97% of the people of Jordan by 1996, and connection to water networks was available to all, rich and poor. Wastewater collection and treatment plants were erected for environmental and public health reasons.

Accessibility to drinking water favorably impacted women and young people, who formerly had to haul drinking water from nearby or remote springs, forcing them to miss school to perform such chores. Access to wastewater collection networks had a positive financial impact on the beneficiaries, reduced public health hazards, and enhanced the environment. Water quotas induced by shortages negatively affected poorer neighborhoods, as people were obliged to purchase water from vendors at much higher prices than they would pay for network water. Irrigation-water quotas gave priority to orchards and groves, with less shares to open-field crops, thus having a built-in bias toward the rich.

Issues of allocation usually entail bias for urban areas at the expense of rural areas. Agricultural water has the advantage of balancing population distribution between rural and urban areas, generates jobs at cheaper capital cost per job, and has a positive environmental impact. Agriculture can use treated wastewater, whereas most other sectors cannot. Industrial allocations command higher priority than agriculture, and environmental water started to receive attention.

Some inequity issues in municipal water allocation and accessibility are mitigated by cross-subsidies inherent in the water tariff, and some inequities in allocation to agriculture are mitigated by the government subsidies accorded to it.

Policy on treated wastewater reuse in agriculture was prompted by water stress and the relative position of the consumption centers in relation to the Jordan Valley. This made it possible to pump agricultural blue water from the valley for municipal uses in Amman and replace it with gray water from the city. Issues of inequity were raised by valley farmers, as this replacement had adverse economic and environmental impacts.

The integrated social and economic development of the Jordan Valley is analyzed to demonstrate the various social and cultural impacts of water use as a backbone of such development. Data cited date back to 1986, when the majority of integrated development projects were completed. Unfortunately, no subsequent study was made to evaluate these impacts more recently.

The integrated development succeeded in reversing the trend of migration from rural to urban areas. It resulted in economic growth of the valley with structural changes, along with parallel changes in regional employment distribution in favor of female participation in the labor force. The building of social infrastructure resulted in much improved education and health services, whose indicators matched or surpassed the national averages. Housing indicators greatly improved. The impact on social dynamics was pronounced. Society in the valley became more material and outward-looking, moving away from the traditional tribal values. Traditional marriage structure changed. Enrollment of females in schools increased drastically, and women became an integral part of the workforce like never before. Attitudes and expectations became more optimistic. Among the positive indicators of life quality changes, life expectancy at birth increased and became 5% higher than the national average, and the incidence of infant mortality and child morbidity were reduced.

Society in the Jordan Valley ascended to active participation in the political life of the country. Valley residents were elected to the national Parliament, and some were appointed to the Senate. Others assumed portfolios in the cabinet. Nothing like this was enjoyed by valley residents before integrated development was implemented.

A unit flow of water (1 mcm/y) is capable of creating an opportunity for about 150 families to settle in rural areas. If that unit flow is diverted to urban uses, these families will follow the water in search of jobs and a livelihood. The average cost is less for an agricultural job than for a job in industry.

The pros and cons of the water sector adjustment programs are discussed as they impacted water and agriculture.

References

Abu Yusif Ya'koub Bin Ibrahim Bin Habib Al Ansari. Died 804 CE. *Kitab Al Kharaj*, section on canals, wells, and rivers, 94–97. Beirut, Lebanon: Dar Al Ma'rifah for Printing and Publishing. Reprinted from an original manuscript, no. 674, and verified with the copy printed by Boulaq Press, Egypt, 1924.

Al Rai (Amman, Jordan). 2004. Saudi Delegation in Amman to Discuss Renewal of Import of Jordanian Vegetables. October 7.

CATAD (Centre for Advanced Training in Agricultural and Rural Development). 1996. *Qualitative Impact Monitoring of Agricultural Structural Adjustment in Jordan.* Amman and Berlin: Humboldt University.

DOS (Department of Statistics). 2005. *Statistical Yearbook.* Amman, Jordan: DOS.

Othman, M. 1983. *Water and the Process of Development in the Kingdom of Saudi Arabia.* Jedda, Saudi Arabia: Tihama Publishers.

Shepley, S.C., M. Bitoun, D. Gaiser, and H. Nassif. 1987. *The Jordan Valley Dynamic Transformation, 1. 973–1986.* Study prepared by Tech International, Inc., in association with Louis Berger International, Inc., under AID contract no. ANE-0260-C-00-7054-00.

Notes

1. Wadoo' refers to washing of the limbs, face, and head, mandatory before performing the Muslim prayers. The letter *H* after the date stands for Hijrah, the "migration" of the Prophet Mohammad from Mecca to Madeenah, both in Hijaz, a province of today's Saudi Arabia. The year of the migration coincided with the year 622 AD and constitutes the beginning of the lunar Hijrah calendar.

2. Jordan had been under emergency law between 1957 and 1990, under which the prime minister had been the military governor. The cholera outbreak was such a national emergency that the prime minister used his military powers to manage the situation.

3. The idea is credited to Engineer Omar Abdallah Dokhgan, the president and chairman of the Board of the Jordan Valley Authority. He was assisted by his deputy, Munther J. Haddadin.

4. The meeting was attended by the crown prince, the prime minister, the chief of the Royal Court, the minister of municipal affairs, the president of the National Planning Council, the president and the vice president of the JVA, the vice president and the director general of the Natural Resources Authority, and the mayor of Amman.

5. The states of the Gulf Cooperation Council are the primary importers of the Jordan Valley's produce. They all were unhappy with the stand that Jordan took with respect to the occupation of Kuwait.

6. The northernmost part of the Jordan Valley received rainfall enough to support a limited variety of winter crops.

7. The JVC was created via law number 2 for the year 1973; the JVA was created by law number 18 for the year 1977.

8. Calculated at the current expenditure of US$260 million to irrigate 26,800 ha; each 1.5 ha generate two jobs in direct agriculture.

10

Conflict Resolution and Regional Cooperation

Munther J. Haddadin

The Middle East, of which Jordan is a part, has been the scene of political, social, and cultural conflicts since the victorious Allies of World War I supported the goals of the World Zionist Organization (WZO) to establish in Palestine a national home for the Jewish people. Attempts were made in 1937 and 1938, and again in 1947, to partition Palestine into two states: one for the Palestinian Arabs and another for the Jewish people. None of these attempts succeeded.

Hostilities arose between Palestinians and the immigrants and intensified in the wake of UN Resolution 181 in 1947 to partition Palestine. Waves of Palestinian refugees took to the roads in search of safety in the West Bank and Gaza. War broke out between the army of the newly proclaimed State of Israel and the armies of the surrounding Arab countries, which barely matched the Israelis at the outset and were outnumbered 3:1 as the war progressed (Shlaim 2000).[1]

Another wave of Palestinian refugees took to the roads, ending up in the West Bank, Gaza, and the neighboring Arab states of Jordan, Syria, Lebanon, and Egypt. Bitterness and hatred filled the atmosphere of the Middle East, with Arabs against Israelis, Arab citizens blaming their rulers, and a series of military coups that shaped the style of governments in the Arab countries neighboring Jordan.

The West Bank, a part of Palestine retained by the Jordan Arab Legion, united with the Hashemite Kingdom in 1950 and became part of the Hashemite Kingdom of Jordan. It was against the above background that the water conflict had been building up since 1939.

Competition over Water Resources

The government of Transjordan (today's Hashemite Kingdom of Jordan) found itself in the middle of the conflict. The royal British Peel Commission, the first to recommend the partition of Palestine, further recommended that the Arab part of Palestine, as outlined in their report, be incorporated in the Emirate of Transjordan. To assess the adequacy of water resources, the government of Transjordan commissioned a British engineer, Michael G. Ionides, to study the water resources of the Jordan Basin and determine their adequacy to support the resettlement of population in the context of creating two states as recommended. Three factors are important to highlight:

- The indigenous human resources of the Emirate of Transjordan were not capable of conducting such a study, and the government thus had to rely on expatriate knowledge. This continued for decades afterward.
- The events in the region drove the water agenda for Jordan. Without the Peel Commission report, Jordan most likely would not have bothered with the Ionides study.
- The financial resources of the emirate (and later the Hashemite Kingdom) did not allow large capital expenditures, and thus Jordan had to rely on grants and loans from the outside.

Ionides concluded (1939) that the water resources would be a material constraint; they were not sufficient to support a Jewish state that would be the destination of Jewish immigration in addition to the Arab state envisioned in the Peel Commission's report. Ionides's study estimated, for the first time, the available water resources and irrigable land in the Jordan Valley. His development plan was based on water availability from the Jordan River and Lake Tiberias.

On the part of the WZO, the Jewish agency commissioned an American engineer, Walter Clay Lowdermilk, who was working with the United States Department of Agriculture, to assess the adequacy of the water resources in Palestine to support Jewish immigration. His plan, published in 1944, reinforced Jewish arguments that proper water management would make the water resources sufficient to resettle 4 million Jewish immigrants in addition to the 1.8 million Palestinians and Jews already living in Palestine (Lowdermilk 1944). His recommendations entailed the transport of Jordan River water southward to the dry Negev to make room for Jewish settlement there. He included the Litani, a purely Lebanese river, in his regional management scheme. Lowdermilk's plan called for the transport of seawater from the Mediterranean to the Dead Sea to compensate it for the loss of Jordan River water, an idea rooted in the works of Jewish engineers in the 1870s and taken up by Theodore Hertzl, the founding father of the WZO, and by other engineers later, the most prominent of whom was Simcha Blass (Blass 1973).[2]

The above two plans by Ionides and Lowdermilk constituted the water-planning platforms of Jordan and Israel, respectively.[3] The parties competed for the waters of the Jordan Basin, and two other riparian parties, Lebanon and Syria, joined in the competition, each wanting to utilize as much of the basin's water as it possibly could.

After the State of Israel was proclaimed on May 14, 1948, the influx of Palestinian refugees into Jordan prompted the government to plan for the development of Jordan Valley water resources to create jobs for an expanded population and provide for a basic livelihood. Jordan, proclaimed a kingdom in 1946, was concerned about Israeli water plans and fearful that they would usurp Jordanian waters. It therefore commissioned a British consulting firm, Sir Murdoch MacDonald and Partners, to study economic, social, and political aspects of the development of Jordan Valley water resources. Jordan's policy aimed at maximization of its water rights, assessment of the grounds on which Israeli transfer of Jordan River water outside the basin (as recommended by Lowdermilk) could be counterargued, and planning for utilization of its share of the waters.

The basic principle of the plan formulated by MacDonald and Partners for the Hashemite Kingdom in 1951 was adhered to by the Arab plans that followed. The firm asserted that "the general principle, which to our mind has an undoubted moral and natural basis, is that the waters in a catchment area should not be diverted outside this area unless the requirements of all those who use or genuinely intend to use the waters within the area have been satisfied" (MacDonald and Partners 1951). The Arab states endorsed this principle, but Israel adamantly opposed it because it would abort the Israeli plans inherent in the work of Lowdermilk, as well as subsequent plans elaborating it by an American engineer and consultant for the Jewish Agency (Hays 1948) to transfer Jordan water outside its basin to the Negev. The priority assigned to the in-basin use of the waters of the Jordan River constituted a main source of contention between the Arab riparian parties and Israel.

Another point that proved abortive to Jordanian efforts for the development of the basin's water resources was the imperative of the "inclusion" approach stipulated in several conferences and required by the World Bank. It called for notifying the co-riparian parties on an international watercourse of, and including them in, any plans for development of its waters. Jordan, like all other Arab states, could not include Israel in its efforts because of the hostilities and nonrecognition by the Arab states of the State of Israel; someone had to do it indirectly, and the United States assumed that role.

Riparian Protests over Development Efforts

The already marginal economy of Jordan in 1948 was further burdened with the influx of Palestinian refugees. The country had to rely on exogenous resources, both financial and human, to help in its economic and social development. In the field of water resources, Jordanian know-how started to develop in the early 1950s, with engineering graduates returning from universities in adjacent countries, primarily Iraq and Egypt. No higher-learning institutions existed in Jordan until 1962, when the University of Jordan was established, and it was not until 1978 that the first batch of engineering graduates, admitted in 1973, entered the labor market.

The United Nations Relief and Works Agency for Palestine Refugees (UNRWA), established in 1949, moved in to help Jordan adjust to the new demographic pressure. A reservoir site on the Yarmouk River, the largest tributary to the Jordan, was accidentally observed by an American engineer, Mills E. Bunger, while on a flight from Amman to Beirut, Lebanon. Bunger was working for the Technical Cooperation Agency (TCA) of the United States in Jordan. Building a dam and expanding irrigation in the Jordan Valley would create badly needed job opportunities and help settle Palestinian refugees in the Jordan Valley, where some of them had already settled while waiting for their repatriation to Palestine.[4] Soon the TCA joined in and made funds available (US$929,000), along with the funds put up by the UNRWA (US$856,000) and the Jordan government (US$200,000), to finance the feasibility study of a dam on the Yarmouk at Maqarin, including site investigation work.[5] The Jordan government commissioned an American engineering consulting firm to perform the feasibility studies.[6]

When the Bunger Plan, named after its producer, was first conceived in 1951, it was stipulated that it "should aim at the maximum development of the Jordan Valley without involving international negotiations which might not be feasible at the present moment, and, secondly, should comprise a scheme which could easily be fitted into any subsequent scheme derived from the use of Lake Tiberias" (Welling 1952).[7]

The Jordan government adopted the work of Bunger as policy and recognized the importance of liaison with Syria, the upper riparian on the Yarmouk, to implement that policy. Of the several investigators of the basin's water, Bunger was the first to suggest impounding Yarmouk floods by a dam on the river itself. All the previous investigators suggested the use of Lake Tiberias as a storage reservoir, with a diversion structure across the Yarmouk near Adassiyya to divert its flow to the lake. Bunger did not ignore the previous studies and suggested that the dam could be "easily fitted into any subsequent scheme derived from the use of Lake Tiberias." Subsequent studies adopted Bunger's suggestion and recommended a dam on the river.

Concurrently with the feasibility study for the Bunger Plan, also known as the Yarmouk–Jordan Valley Plan, Jordan approached Syria to work out a bilateral agreement over the Yarmouk, the river that more or less forms the boundary between the two countries.[8] A bilateral treaty was concluded between the two countries on June 4, 1953, for the utilization of the Yarmouk waters.

Syria retained the right to the full use of the springs in its territories above elevation 250 m above sea level and would benefit from 75% of the power generated by the Maqarin (Bunger) Dam. The Maqarin site was more convenient to Jordan than was Lake Tiberias for two reasons: it would enable power generation, and it would store Yarmouk water on Arab lands, whereas the lake was currently under the control of the enemy.

After the studies of the Maqarin Dam were completed, the UNRWA, in agreement with the Jordan government, on March 30, 1953, allocated US$40 million from its rehabilitation funds for the construction of the dam. The total cost of the dam and irrigation works was US$70 million. In July 1953, the UNRWA, TCA, and the Jordan government appropriated funds for the initial expenditures, and the recruitment of workers began (Stevens 1956). Thus the Jordan policy appealed to both the UNRWA and the U.S. administration, as both appropriated funds for its implementation.

It was at this point that the plan was suddenly stalled, to the dismay of the Jordanian government. The U.S. government, which had given its official endorsement, now proposed another approach and a different plan. It turned out that Israel, the third riparian on the Yarmouk and a riparian on the Jordan, had protested to the United States and the United Nations against the Bunger Plan, Jordan's official policy. Israel, in competition with Jordan over water, and in an attempt to initiate direct contacts and talks with Jordan, claimed that the Bunger Plan would jeopardize its rights and demanded that it be included in any planning involving the development of the Yarmouk waters. The implementation of the Bunger Plan was thus aborted, but the Jordan government policy did not change.

Concurrent with Jordan's efforts to go ahead with the Bunger Plan, Israel was proceeding with the implementation of its All Israel Plan (1951–1955), particularly its grand project of water transfer from the Jordan to the Negev. Work on the project started with the Auja–Yarqon to Negev segment. When Israel attempted to start work on the intake structure on the Jordan River immediately south of Jisr Banat Ya'coub in the Middle Demilitarized Zone with Syria, it was met with Syrian opposition. This took the form of protests and complaints to the United Nations Truce Supervision Organization (UNTSO) and the UN Security Council and, occasionally, shelling from Syrian positions stationed on the overlooking Golan Heights. Israel's attempts to divert the Jordan coincided with another water project it was implementing north of and adjacent to the intake structure: the draining of the Huleh marshes. Israel's works were suspended by a decision of the Security Council and, more importantly, a supporting reaction from the United States as expressed by its secretary of state, John Foster Dulles, who threatened to stop aid to Israel if it did not abide by the Security Council resolution.

Thus unilateral plans made by Jordan and by Israel for the development of water resources did not take off because of the protests filed by Israel against

Jordan's plan and because of Syrian opposition to Israel's action to divert waters from the Jordan River outside its basin, expressed sometimes by military clashes. The situation on the Jordan–Israeli lines was no quieter.

In the midst of the onset of the cold war, the United States, the uncontested leader of the free world, decided to mediate among the riparian parties and proposed the Unified Plan for the development of the Jordan Valley. Such a plan would fight want, a breeding environment for communism; help resettle Palestinian refugees in the Jordan Valley; and pave the way to acceptance of Israel by its Arab neighbors.

Diplomacy over Shared Water Resources

Parallel to the slowdown of American and UNRWA support for the Jordanian Bunger Plan, the UNRWA, upon the recommendation of the British Foreign Office and with the tacit approval and participation of the United States, quietly moved to propose a plan for the sharing and utilization of the waters of the Jordan River Basin.[9] The projects resulting from a sharing plan would assist in alleviating the difficulties facing the Palestinian refugees, especially in Jordan, by creating viable means of living for them in the Jordan Valley. Additionally, a shared plan would ensure coordination, if not cooperation, among the riparian parties in the basin and prevent one riparian from inflicting damages on another, willfully or accidentally, by implementing projects using the same water that a co-riparian was already using.

The UNRWA asked the Tennessee Valley Authority (TVA) to conduct a study of a plan for the development of the Jordan Basin in the territories of all four riparian states.[10] The TVA entrusted that task to a specialized American consulting firm, Chas. T. Main, commissioning it to do a desk study and formulate the plan.[11] The TVA–Main Plan was submitted to the UNRWA and the U.S. government on August 31, 1953, and to the concerned Arab states a month after that date.

President Dwight D. Eisenhower tried to diffuse the tension in the strategically sensitive area of the Middle East, especially the clashes caused by water diversion. Eisenhower dispatched to the region a presidential envoy, Ambassador Eric Johnston, with the mission of getting all the riparian parties on the Jordan to subscribe to a negotiated unified plan for the development of the Jordan Valley.[12] Ambassador Johnston arrived in Beirut, Lebanon, on October 22, 1953, in his first stop in the region. He presented the TVA–Main Plan to Lebanon's prime minister, Abdallah al Yafi, and visited Jordan, Egypt, and Syria, where he presented the same plan and asked that the parties study it and respond to it.

Jordan's policy stood against a unilateral diversion by Israel of the water of the Jordan River. The policy stemmed not only from an intention to deny Israel water it would use in the Negev to house more Jewish immigrants, but also from Jordan's reliance on the fresh water of the river for the irrigation of about 6,800 ha in the river's floodplain, known as the Zor, which was divided almost equally between the West and East Banks. Private farmers had installed pumps and dug up irrigation canals to develop those areas. Diversion of the freshwater resources would threaten the sustainability of farming there.

Another component of Jordan's policy in regard to Israel's diversion was the kingdom's standing position not to violate any of the decisions of the League of Arab States that were opposed to any accommodation of Israeli interests prior to the resolution of the major issues connected with the establishment of the State of Israel, most notably the issue of Palestinian refugees. A corollary of this policy was to support any and all Arab brethren in their confrontation with Israel; Syria was engaged in countering the Israeli diversion project, and so was Jordan. Additionally, an influential portion of Jordan's population had its source and origin in Syria, and thus cordial brotherly relations with Syria were among the objectives Jordan pursued as much as politically possible.

Jordan's policy was not to act unilaterally on matters that concerned more Arab riparian parties than just Jordan; it viewed coordinating with them as absolutely necessary. Egypt was looked upon as the senior Arab party, with qualifications to lead Arab efforts in that regard. Home to the League of Arab States, Egypt was run by a revolutionary council that was gaining popularity in Jordan and the Arab World. Egypt's strong man, Jamal Abdul Nasser, formed an Egyptian team in the wake of Johnston's visit to study the plan he had presented. Nasser soon decided that the matter should be looked into by the League of Arab States, and not just by Egypt alone. The league, influenced heavily by Egypt, formed a technical committee headed by a qualified Egyptian engineer and empowered it to study the plan and discuss it with Ambassador Eric Johnston. No representative of Jordan was appointed in the committee, but that was rectified four months later.

Jordan was careful not to antagonize Egypt or any other Arab party. The young King Hussein had assumed constitutional powers shortly before Johnston arrived in the region. Additionally, the Arab states were upset with Jordan on account of its unity with the West Bank in 1950; they considered such unity an annexation that would dilute the Palestinian entity. No Arab state recognized that unity. Jordan was careful not to cause further antagonism to any Arab state by going it alone on the matters concerning the Jordan Basin, or any other matter that concerned more Arab states than just Jordan.

Johnston conducted four rounds of talks through shuttle diplomacy, in which he negotiated separately with the Arab Technical Committee and with Israel. This culminated in what Johnston called the Unified Development Plan for the Jordan Valley. In it, he spelled out the shares of each riparian party and described the structures that would be needed to implement the plan. The Arab Technical Committee, though not wholeheartedly, recommended to the

Arab League Council that it accept the plan, and the Israeli side, headed by Minister of Finance Levi Eshkol, conceded to it. However, the Arab League Council decided, for political reasons, to defer approval of the plan until further studies were made, and the Israelis had a disagreement with Johnston over their final share from the Yarmouk River. The plan recommended the use of Lake Tiberias as a common reservoir for Jordan and Israel to regulate Jordan River and Yarmouk River waters. It also deferred Yarmouk storage in the lake for five years and left the opportunity open to find another economic site in which the Yarmouk waters could be stored.[13]

In brief, the shares of the riparian parties per year were stipulated in the plan as follows:

- Total flow of the Jordan River: 903 mcm/y (157 mcm from Hasbani, 258 mcm from Dan, 157 mcm from Banyas, 68 mcm intermediate flow, 65 mcm Huleh drainage, and 198 mcm direct inflow into Tiberias). To this were added 86 mcm of return flow and 27 mcm of historic consumptive use (HCU) by irrigated agriculture in the basin before the plan was formulated. The total usable flow was 1,016 mcm. A ceiling of 15 mcm of brackish water was added to the share of the West Bank. The grand total was 1,031 mcm.

Shares from the Jordan:

- 35 mcm for Lebanon from the Hasbani tributary (3 mcm HCU plus 32 mcm new allocations, of which 12 mcm would return to the river).
- 20 mcm for Syria from the Banyas tributary (4 mcm HCU plus 16 mcm new allocations, of which 8 mcm would return).
- 22 mcm for Syria from the Jordan River course (10 mcm HCU plus 12 mcm new allocations, of which 5 mcm would return).
- 100 mcm for Jordan (West Bank) to be drawn from Lake Tiberias, with a ceiling of 15 mcm of brackish water; no usable return flow.
- 300 mcm allotted to evaporation from Lake Tiberias.

The remainder, estimated at 554 mcm of the flow, including 65 mcm gains from the drainage of Huleh swamps and 86 mcm of return flow, was allocated to Israel.

- Total flow of the Yarmouk River: 506 mcm/y (467 mcm measured at Adassiyya; 13 mcm HCU; and 26 mcm return flow from Syrian new allocations).

Shares from the Yarmouk:

- 90 mcm for Syria (5 mcm HCU plus 85 mcm new allocations, of which 26 mcm would return).
- 25 mcm for Israel (no return flow).
- 14 mcm allotted to evaporation.

TABLE 10-1. Jordan's Water Shares in the Basin as per Johnston's Allocations (mcm/y)

Water source	East Bank	West Bank	Total
Side wadis	175	52	227
Groundwater wells	8	8	16
Yarmouk River	296	81	377[b]
Jordan River	0	100[a]	100
Total	479	241	720

a. To include a ceiling of 15 mcm of brackish water from springs around Lake Tiberias.
b. Based on an exaggerated 26 mcm/y of return flow from Syrian new allocated uses of 85 mcm/y.

The remainder, estimated at 377 mcm of the flow, was allocated to Jordan (296 mcm for the East Bank and 81 mcm for the West Bank).

The side wadis inside Jordan were considered part of Jordan's share in the East Bank (175 mcm) and West Bank (52 mcm). Additionally, groundwater abstracted from wells in the east Jordan Valley (8 mcm) and west Jordan Valley (8 mcm) was counted among the shares of the Hashemite Kingdom of Jordan. In effect, Jordan's water shares in the basin were stipulated as shown in Table 10-1.

It has been subsequently noted that the return flow from Syrian uses in the Yarmouk Basin, initially estimated at 26 mcm out of 85 mcm of new allocations to Syria, or about 30%, has been on the high side, and that it is more likely about 10%.[14]

Jordan's Policy toward the Unified (Johnston) Plan

The Arab League communicated to Ambassador Johnston on October 15, 1955, its desire to defer decision on the Unified Plan pending more studies to be performed by the Arab side. The League Council did not reject the plan. Political developments that followed pressed Jordan for accelerated economic development to create jobs for an increasing number of entrants in the labor market. Under the Johnston formula, each riparian party would be free to use its share of the water the way it saw fit, which meant that Israel could divert from its own share any part that it desired to wherever it wished. Calculations of water shares made by the Johnston mission indicated that an annual flow of 320 mcm was allotted to Israel's grand project, the National Water Carrier, to be consumed in the Negev Desert outside the Jordan Basin.

To avoid more political complications, Lebanon asked the United States to seek Israel's commitment not to divert Jordan River water until agreement on its use could be reached.[15] It was realized that an indefinite postponement would not be acceptable to Israel. Lebanon's foreign minister, Salim Lahoud, expressed the possibility of implementing the Jordan Valley Unified Plan by the Arabs and, separately, by Israel. Lahoud asked that the Johnston team prepare a modest plan of limited operations that could be readily implemented and have major projects in the Unified Plan postponed until consensus about them was reached. Projects given as examples of limited operations were a diversion dam on the Yarmouk at Adassiyya, partial development of the East Ghor in Jordan, expansion of Buteiha Farm in Syria, and field studies of the Hasbani in Lebanon. This proposal by Lebanon's foreign minister opened the way for conception of projects on both the Arab and Israeli sides suitable for a staged implementation of the Unified Plan.

Jordan, as a matter of policy, never violated any resolution adopted by the League of Arab States, nor has it ventured into projects or activities that could be construed as direct or indirect cooperation with Israel. Under such policy, Jordan accepted the recommendations of the Arab Technical Committee on the Johnston Plan, even though Jordan's water shares were being diminished in every round of talks Johnston made with the committee. Because the Arab League Council did not readily endorse the Johnston Plan, Jordan did not either. Consensus in Arab decisions was almost impossible to reach in the wake of the Arab "cold war," which commenced in 1957 and lasted until the eve of the June war of 1967.[16]

The East Ghor Canal Project

Jordan's East Ghor Canal Project was rooted in the notion advanced by Minister Lahoud of Lebanon to implement partial and separate projects under the Unified Plan. Almost simultaneously, Israel conceived the Beit Shean Irrigation Project, meant to irrigate the Israeli side of the Jordan River Valley with Jordan River water drawn from Lake Tiberias.

Jordan did not have water engineers at that time; the Jordanians who participated in the work of the Arab Technical Committee in its negotiations with Johnston were a surveyor and a politician–economist. Outside technical assistance was needed to look into the water projects of the Unified Plan. Jordan's Ministry of National Economy asked the UN Food and Agriculture Organization (FAO) to look into a proposal to undertake a pilot scheme for the development of the Jordan Valley in preparation for partial implementation of the Unified Plan. The FAO dispatched an expert, S.N. Simansky, to study that proposal, come up with a plan to draw off some of the base flow of the Yarmouk River, and develop a pilot scheme. Simansky studied three variants, and Jordan embarked on one that would allow staged construction of the East Ghor Canal and distribution networks to irrigate a total of 10,000 ha with Yarmouk flow.

The Jordan government approached the government of the United States for assistance in implementing the chosen variant. The International Cooperation Agency (ICA) of the United States, the successor of the TCA and predecessor of the United States Agency for International Development (USAID), expressed readiness in 1958 to finance the project and assisted in retaining a consulting engineering firm.[17] The ICA's interest in financing Jordan's project came after a comprehensive bilateral assistance umbrella agreement was signed between Jordan and the United States in 1957, pursuant to the termination of the Jordanian–British Treaty and after substantial turmoil in the region in response to the conclusion of the Baghdad Pact and the Suez Campaign.[18]

From a policy viewpoint, Jordan's economic situation in 1957 could hardly have been less enviable.[19] It had an estimated population of 1.6 million and was burdened with Palestinian refugees, unemployment, and the need for government military spending in light of Israel's repeated assaults on Jordan's frontiers in the West Bank. The country was severely limited in natural resources, technical capacity, and transport facilities and was beset by internal and external political pressures. During the 10 years since its proclamation as an independent kingdom in 1946, Jordan had been dependent on foreign aid for more than half its annual government budget and more than 60% of its total revenues. In 1957, the deficit in the balance of payment was the highest of any country in the Middle East except Israel, while the per capita share of the annual income, at US$100, was the lowest. Agriculture constituted about 46% of its gross domestic product (GDP), and employment depended on rain-fed farming and livestock.

In light of the political and economic rift developing with its neighboring Arab states, it became imperative for Jordan to launch development projects, create new jobs, and assuage its citizens' political frustration, exacerbated by hostile broadcasts from Damascus and Cairo. Foreign financial and technical assistance was badly needed to launch development projects, and the United States responded to Jordan's needs. In January 1958, the U.S. National Security Council authorized, in paragraph 40 of resolution 5001, a piecemeal approach to the implementation of the Johnston (Unified) Plan.

Jordan intended to have the pilot scheme implemented in stages as well. It was a project that did not require too much training before people could reap the benefits; Jordanians (and Palestinians) have been familiar with farming and agriculture since the dawn of human civilization; the project would put to use the kingdom's rightful shares of the Yarmouk waters, generate jobs in agriculture and supporting services, and enable the development of a community in the Jordan Valley that was capable of production and would protect the valley against Israeli ambitions. A staged implementation would enable Jordan to cope with the supervision and management of its first large project and build indigenous technical and managerial capacity, and it would facilitate the flow and spending of funds.

Before an agreement over the project could be concluded with the United States, Jordan had to issue a commitment that it would not use more of the Yarmouk River water than allocated by the Unified Plan. This was done by Jordan's minister of foreign affairs, Samir al Rifai, in memorandum number

58/14/6719, dated February 25, 1958, addressed to the U.S. chargé d'affaires in Amman. The note acknowledged the chargé's statement made the day before in the following aide memoire from the U.S. embassy:

> The American Chargé d'Affaires called on Foreign Minister Rifa'i this morning and informed him that the United States Government had instructed him to state that it is prepared to assist Jordan in financing the first year costs of the proposed Yarmouk Diversion Project and the East Canal. It is expected that this financing will be accomplished by use of fiscal year 1958 funds already allotted for economic development in Jordan. The Chargé said that he had also been instructed to inform the Foreign Minister that this assistance will be extended provided that there is an explicit undertaking from the Hashemite Kingdom of Jordan that it will not draw from the Yarmouk River more water than allotted to it under the Unified Development Plan.

Conforming to the division of water as stipulated in the plan has evidently become the policy of the Jordan government regarding the international waters of the Jordan River Basin. This policy and commitment of the Jordan government continued to guide the government officials who were in charge of the development of the Jordan Valley and the operation of its waterworks. Between 1959 and the outbreak of the June war of 1967, the Jordan government, aided by grants from the United States, brought under irrigation a total area of 114,000 du (11,400 ha) via a network of open canals fed by a 70-km-long main canal drawing water from the Yarmouk River. The diversion of water from the Yarmouk River to the canal was done by a drop inlet; the construction of a diversion weir to midstream, as called for in the original pilot scheme, was not possible because of Israeli objections.

Water Policy toward Israel

The implementation of the pilot scheme would eventually reduce the contribution of the Yarmouk River to the Jordan River flow. Israel depended on both the Jordan and the Yarmouk for irrigation of lands in the basin. Jordan relied on the Jordan River for the irrigation of private lands along the riverbanks.

Obtaining U.S. approval and finance for the pilot scheme benefited Jordan in several ways. It made possible the start of development of the Jordan Valley and the realization of objectives listed above. The finance was in the form of grants, a transfer of foreign currency that did not have to be repaid. And important politically, it brought in an interested third party to deal with Israel. Jordan could not make any contacts with Israel, with which it was at war, and which Jordan and the other Arab countries had never recognized in the first place.

Israel was dismayed that the United States had not consulted with it before approving the Jordanian project and forwarded notes expressing worries that it might cause appreciable harm to Israel. In its notes, Israel requested assurances from the United States that ways and means be found to safeguard its established rights, as defined in the Unified Plan, and mentioned that Israel's share in the Yarmouk waters as per that plan was 40 mcm/y.[20] An attachment to the Israeli note went into technical details to show that there would not be enough water in the summer months to support the Jordanian project, except at the expense of the Israeli share. The United States was firm on these points and assured the Israelis in a memorandum dated August 1, 1958, that maximum diversion by Jordan would not exceed 70% of the summer flow, or about 5.1 cubic meters per second (cms), leaving adequate flow for Israeli use, and that the East Ghor Canal Project was within the provisions of the Unified Plan.

Parallel Israeli Projects

The competition over water resources and the conflict among the parties over their shares prompted each co-riparian to advance its own plans to utilize its shares. When Jordan succeeded in receiving U.S. support to build the East Ghor Canal, Israel seized the opportunity to request similar treatment from the United States. It presented two projects for U.S. consideration: the Tiberias–Beit Shean Project, which would ensure the sustainability of irrigated lands in the Beit Shean area after Jordan diverted flows from the largest tributary, the Yarmouk, and would expand irrigation there; and the National Water Carrier, which was capable of transferring 425 mcm of Jordan water from Lake Tiberias to the Negev, as opposed to the 320 mcm envisaged in Johnston's calculations. Israel got U.S. financial support for the Beit Shean Project, but the United States hesitated to back the National Water Carrier, which the Arab parties resented. Later, in 1961, the United States went through a cutoff agency, the Agricultural Credit Bank in Israel, to supply financing for the remaining parts of the carrier project.

When Israeli efforts in building the National Water Carrier intensified and were close to completion in 1964, Nasser of Egypt called for an Arab Summit Conference in Cairo to discuss the matter. The Arab League approved the Arab Diversion Project, which was capable of diverting portions of the Banyas, Hasbani, and Yarmouk tributaries. A special organization was set up for the implementation (Chapter 2), but the efforts were stopped right after the June war of 1967. The Arab Diversion Project could have secured the Arab shares, as stipulated in the Unified Plan, without negotiations or cooperation with Israel.

Jordan's Water Diversion Policy

The East Ghor Canal Project started its operations in 1962, and water was diverted from the Yarmouk via a lateral drop inlet into a 1-km-long tunnel designed to pass 20 cms. When the tunnel saw daylight near Adassiyya, the main canal started, with a conveyance capacity of 10 cms, raisable to 20 cms. The capacity was raised in 1966, when it was decided to build a dam on the Yarmouk at Mukheiba within the Arab Diversion Project approved at the Arab Summit in September 1964.

Jordan is the middle riparian party on the Yarmouk, and Israel is the middle riparian party on the Jordan. The upstream riparian party on the Yarmouk is Syria, and on the Jordan the upper riparian parties are Lebanon and Syria. Because of Israel's superior military, economic, and political power, Jordan has always found its position on the Yarmouk tenuous, as if it were between a rock and a hard place. It could not jeopardize Israel's rightful water share, by virtue of the commitment Jordan made to the United States and by virtue of being a weaker military power compared with the other two riparian parties, and it could do nothing if Syria were to violate the shares allotted to it under the Unified Plan. As a matter of fact, the Syrian share as stipulated in the Unified Plan was the same as Syrian members of the Arab Technical Committee asked for during the negotiations with Ambassador Johnston, and it was equal to the integrated flow of the springs above elevation 250 m above sea level located in Syrian territories in the Yarmouk Basin. This was fully compatible with the Syrian share stipulated in the bilateral Jordanian–Syrian treaty over the Yarmouk concluded on June 4, 1953, in which Syria retained the right to use the water of the springs emerging in its territories above elevation 250 m above sea level.

Syria had a historic consumptive use on the Yarmouk in the order of 5 mcm/y and was gradually increasing its uses of the springs for production of summer crops in the eastern Golan and western Horan regions. The historic consumptive use of lands in the Yarmouk Triangle that ended up in 1948 in Israeli hands was 10 mcm, and Israel was expanding its agriculture in the Yarmouk Triangle. Jordan started utilizing water from the Yarmouk in 1962 and was gradually expanding under the staged East Ghor Canal Project. Given these facts and the prevailing political complications, with Jordan and the Arab states being in a state of war with Israel, Jordan's policy on water diversion was to do the following:

- Strictly conform to the provisions of the Unified Plan to honor the commitment issued by the Jordan government to the United States on February 25, 1958.
- Work closely with the United States with genuine transparency, and be clear with the U.S.-based consultants on the project when instructions from their Jordanian employer were issued.

- Welcome visits by officials of the United States to follow up on the project to make sure the funds were used for the purpose for which they were appropriated and to see the progress made. Dr. Wayne Criddle, the technical arm of Johnston's mission, in particular made important visits to the Jordanian and Israeli sides to make sure neither violated the Unified Plan in the course of implementating the projects.
- Lean on the U.S. diplomatic mission in Jordan to resolve any standing problems that could arise requiring communication with Israel. The United States played that role of intermediary and facilitator to enable progress to be made.
- Permit indirect contact with any Israeli official only to resolve transient problems that could arise on the armistice lines, and only under the auspices of the Armistice Agreement signed between the two countries in April 1949. Military representatives would communicate with each other through a UNTSO officer. No other form of contact was allowed between any Jordanian and any Israeli.

On the Syrian front, Jordan's policy was to adhere to the bilateral agreement signed on June 4, 1953. A joint Jordanian–Syrian committee was established, and it addressed matters of concern to either party. However, the political atmosphere between the two countries was a primary factor in the success of those meetings or lack thereof.

Complications in Policy Implementation

The member behind postponing the endorsement of the Johnston proposals by the Arab League Council in October 1955 was Syria. Its argument was political and focused on the consequences of such endorsement. Syria maintained that endorsing the Unified Plan entailed an indirect recognition of the State of Israel, something the Arab states were not prepared to consider before resolving the standing problems of territorial disputes and the issue of Palestinian refugees. Ambassador Johnston lobbied for the Unified Plan with Prime Minister Jamal Abdul Nasser of Egypt, who pledged to obtain support for the plan before March 1956, the date when Israel would resume work on its own diversion scheme from the Jordan River.

On the Syrian front, Jordan conformed to its 1953 treaty with that nation. The political atmosphere between Jordan and Syria has been strained most of the time since April 1957, when the Arab "cold war" commenced. Their relations went through a hot and cold cycle for more than half a century, and the political thermometer affected the workings of the joint Yarmouk committee to the extent that it never met between 1967 and 1975 or between 1980 and 1986.

On the Israeli front, Jordan had no diplomatic relations with that nation, as the two countries were in a technical state of war. The United States served as a go-between for the two sides, and the provisions of the Armistice Agreement between them served to avoid clashes over water disputes.

In addition to political complications, technical complications arose from a number of causes. First, only annual shares in mcm, and not transient rates in cms, were specified in the Unified Plan, which assumed the full regulation of the Yarmouk flow. In preparation for the East Ghor Canal Project, the ICA had worked out a formula for sharing the transient flow. It stated that 30% of the summer flow, or 2 cms, should be passed to Israel, and that the total quantity per annum should not exceed 25 mcm (or 40 mcm, according to Israel). However, a statement to this effect was never found in the files.[21] Absence of documented evidence of that formula caused Jordan sustained headache for years after 1979.

Prior to the June war of 1967, Jordanian maintenance crews were able to access the Yarmouk River freely, especially at the diversion point, where they cleaned up deposits in the riverbed after each flood season. Israel overran the Golan Heights of Syria in June 1967, including the northern bank of the Yarmouk all the way east to the confluence with Wadi Raqqad. After the Israeli occupation of Syrian territory in the June war, the midstream of the river became a cease-fire line with Jordan (on the southern bank). Jordan's maintenance crews could not do their regular job of cleaning up the riverbed, and year after year such deposits accumulated, eventually forming a sandbar that resembled a narrow but elongated island in the middle of the river. The sandbar affected the efficiency of the diversion to Jordan and reduced the volume during low-flow months. That was evidently to the advantage of Israel.

Other problems involved Syria. Without notifying Jordan of its intentions, Syria started to increase the use of upstream springs emerging from its own territories, which reduced the flow in the river downstream, especially during peak irrigation demand. Syria also expanded its use of the Yarmouk aquifer feeding the lower springs (Jordan's share) and further diminished the Yarmouk flow, especially in the dry months. In 1967, again without alerting Jordan to its intentions, Syria started to build dams on the tributaries to the Yarmouk in its own territories. As such, Syria was putting to use floodwaters earmarked for Jordan to be impounded by the future Maqarin Dam. This act was in clear violation of the 1953 bilateral agreement between the two countries. Jordan, unaware of the Syrian development activities, initiated its integrated development plan for the Jordan Valley in 1973, and by 1979, it had added a total of 83,500 du to the irrigated area, calling for additional diversions from the Yarmouk. This coincided with increased Syrian abstractions from the sources of the river upstream and Israel's increased dependence on water from the Yarmouk that it pumped year-round. Israel's pumping from the Yarmouk approached 75 to 90 mcm/y, compared with its 25 mcm share in the river's water.

At a time when Jordan needed the maximum rate of allowable diversion from the Yarmouk, Syria was working to maximize its abstractions from the springs and the flood flow to Jordan's harm, and Israel would not allow the Jordanian crews to clean the riverbed in order to increase diversion efficiency.

Jordan pursued a policy of direct and friendly contacts with Syria but got nowhere. Syria continued to construct small earth dams on the Yarmouk tributaries and increased its abstraction from the springs at whatever elevation above sea level it could technically reach. It even started to abstract water at elevations of 20 to −50 m below sea level, from the river base flow. Faced with the realities on the ground, and eager to obtain Syria's consent to build the Maqarin Dam on the river, Jordan succumbed to Syrian pressure to rewrite the 1953 agreement. A new agreement was concluded in 1988, by which Jordan conceded to Syria the right to the water impounded by 26 dams in the Yarmouk catchment, and thus reduced the height of the Maqarin Dam from 148 m to about 95 m. It also agreed to give Syria 10 mcm/y of the water impounded by the Maqarin Dam (renamed Wehda Dam) and involve Syrian contractors in the construction activities. The Wehda Dam is now under construction, but Syria never ceased building its small dams.

Jordan relied heavily on U.S. interventions with Israel to have the physical conditions at the diversion site corrected. Jordan was tougher with Israel than it was with Syria, however, most likely because its ties with Syria had been multifaceted and because of demographic linkages. Several other interests bound Jordan and Syria, which was not the case with Israel.

At one point, in July 1979, Jordan mobilized its armed forces to reinstate the water sharing in the Yarmouk that Israel had managed to alter. Israel responded with similar mobilization of the Israeli Defense Forces. Less than 150 m separated the two armies, and I had to manage the confrontation and finally succeeded.[22] However, until the peace treaty was concluded in 1994, the conflict with Israel over the sharing of the Yarmouk River water was managed on basis of ad hoc arrangements through the Armistice Agreement of 1949 under the auspices of UNTSO, with assistance from the U.S. offices.

Regional Cooperation over Water Resources

Jordan became aware of the magnitude of municipal water needs for the majority of its population only in 1979, when a comprehensive study was undertaken by the National Planning Council.[23] The study revealed that by the year 2000, Jordan would face a municipal water shortage of 160 mcm unless more resources were mobilized. This triggered two water policy changes (discussed in more detail in Chapter 4) and added another. The first change in policy was that the water shares in the Yarmouk, heretofore allocated to agriculture in the Jordan Valley, were partially allotted to municipal purposes. Second, municipal wastewater would be treated to standards that would allow its reuse in unrestricted irrigation in the Jordan Valley. Because the consumption centers were located on the plateau, wastewater could flow down by gravity to the Jordan Valley. Another policy component was added in light of the need for regional cooperation to deal with the water shortage. In this regard, Jordan approached Iraq in the hope that it would agree to make up for Jordan's shortfalls in municipal water supplies. Iraq responded favorably to Jordan's request and agreed to supply it with 160 mcm, the projected deficit by the year 2000, from the Euphrates River. Clearly, Jordan is not a riparian party on the Euphrates, and the Iraqi move was bold because of riparian conflicts on that river with Syria and Turkey, and because of the absence of any agreement regulating the riparian relations.

With progress on the Yarmouk stalled, Jordan pursued the Euphrates option and appointed a British consulting firm to perform a study for transporting 160 mcm of Euphrates water to Jordan.[24] The study, completed in 1984, determined that the cost of bulk supply of the Euphrates water to the outskirts of Amman would be about US$2/m^3, plus the cost of distribution and unaccounted-for water. The cost was viewed above the affordability level of Jordanians, and the project was shelved. The two regional cooperation projects, the Yarmouk and the Euphrates, had gotten nowhere despite the good intentions of both Jordan and the Iraqi government.

Intermittent cooperative efforts took place with Egypt and with Syria, especially in the field of advanced irrigation systems that employ drip methods and also in modern agricultural practices. Agricultural technology, such as improved seeds and seedlings and plastic houses, leaked to Jordan from Israel under the open bridges policy Jordan had adopted with the Israeli-occupied territories since 1967. Cooperation with Yemen and with the United Arab Emirates started in the mid-1990s. Groups of farmers and of water officials from Yemen started to visit the kingdom under bilateral cooperation programs. Some training was given to the Yemeni cadres who worked in water and in irrigated agriculture. Gulf states water officials attended specialized seminars held by the Water Authority of Jordan (WAJ) on water management and on operating and maintenance techniques of municipal water systems. More recently, after the U.S.-led occupation of Iraq, Iraqi officials started to receive training at Jordan's Ministry of Water and Irrigation under USAID sponsorship to help them restart their crippled economy after a dozen years of wars and embargo.

It is evident that Jordan could be a center for training professionals from the countries of the Levant and the Gulf. The private sector in Jordan is showing interest in participating in the training and qualification services, and it has been leading in the development of irrigated agriculture, in which it has made investments with encouragement from the Ministry of Water and Irrigation and the Agricultural Credit Corporation.

Other forms of regional cooperation exist in the trade in food commodities. Clearly, food imports conserve indigenous water resources that otherwise would be used to produce the goods. Jordan exports off-season fruits and vegetables grown in the Jordan Valley to Lebanon, Syria, Iraq, and the Gulf states. It also imports food commodities from such regional partners as Syria, Lebanon, Saudi Arabia, Egypt, and Israel, as well as other parts of the world, such as Australia, the United States, and European countries.

An important form of regional cooperation in water resources, and the catalyst in updating water policies, is that between Jordan and the regional funding institutions. The Kuwait Fund for Arab Economic Development contributed substantially to the construction of 8 km of the East Ghor Canal Project, from km 70 to km 78 (1967–1970); the King Talal Dam (1971–1977) and its raising (1984–1988); the largest municipal water supply project, Deir Alla–Amman (1982–1987); and the Safi Irrigation Project, south of the Dead Sea (1982–1986). The Arab Fund for Economic and Social Development helped finance the conversion of some 60,000 du (6,000 ha) from canal surface irrigation to the more efficient pressure pipe networks (1986–1990), as well as the construction of the Karama Dam in the Jordan Valley (1992–1997) and the long-awaited Wehda Dam on the Yarmouk River (2001–2006). The Abu Dhabi Fund provided funding for the construction of the King Talal Dam (1971–1977). The Saudi Fund for Development contributed to the irrigation projects in Feifa and Khneizeera, south of the Dead Sea (1982–1986). Contributions of these funds to Jordan's water sector entailed mutual agreements on certain aspects of water policy, most notably tariffs and administration.

Nonregional financing parties also helped in building water projects. The World Bank and its International Development Association, USAID, Kreditanstalt für Wiederaufbau, the Overseas Economic Cooperation Fund of Japan, the government of Italy, the European Investment Bank, and others contributed substantial funds for the development of Jordan's water sector. Serious debates took place with these funding agencies over the formulation of a water policy, especially in the 1990s.

Water in the Peace Treaty

The peace process that was started in Madrid in October 1991 yielded two conferences: bilateral negotiations between Israel and each of Lebanon, Syria, and a joint Jordanian–Palestinian delegation, and a multilateral conference in which 38 to 42 countries of the world participated. The mission of the bilateral conference was to resolve the political issues between Israel and its counterpart parties, and that of the multilateral conference, functioning through five working groups, was to encourage progress in the bilateral conference and reinforce the peace through regional cooperation projects devised by its working groups on Regional Security and Arms Control, Regional Economic Development, Water, Environment, and Refugees.

An important component of the bilateral peace negotiations was Jordan's policy in negotiations over water. Jordan proposed at the beginning of the negotiations in April 1992 to resolve the conflict over water by falling back on the solution that had been accepted by all parties on the technical level in 1955: the Unified (Johnston) Plan. Jordan continued to rely on that plan as a minimum position below which it would never go, being mindful not to accept less than what the Arab Technical Committee had accepted in 1955 through its negotiation with Ambassador Eric Johnston. Failing agreement on this initiative, Jordan feared that negotiations would have to start all over from scratch.

One major development called for the adjustment of Jordan's share as stipulated in the Johnston Plan. The king of Jordan's royal decree to disengage with the West Bank on July 31, 1988, severed legal and administrative ties with that region, leaving it to the Palestine Liberation Organization (PLO) to deal with the issue of its liberation from Israeli occupation. The water shares that had been allocated to Jordan in the Unified Plan of 1955 pertained to the kingdom at that time, including both the East Bank (Jordan after the disengagement decree) and the West Bank (totally occupied by Israel since 1967). By virtue of the disengagement decree, Jordan had to resolve the conflict with Israel, through negotiations, over the shares of the East Bank only as defined by the Arab Technical Committee. It became the PLO's responsibility to negotiate with Israel over the shares of the West Bank that had been, prior to the disengagement decree, included in the shares of the Hashemite Kingdom of Jordan.

However, under the Letters of Invitation to Madrid, the Palestinians and Israel would resolve their political differences, including the water conflict, in the final status negotiations, slated to begin three years after the commencement of the bilateral negotiations. Accordingly, there would be a time lag between putting water issues on the negotiation agenda between Israel and each of Jordan and the Palestinians. And so it was. Water was listed as the third of eight items on the Jordan–Israeli Common Agenda, but putting the water dispute on the Israeli–Palestinian agenda had to wait for the commencement of the final status negotiations, which have not yet started as of 2006.

As stipulated in the work of the Arab Technical Committee, the water shares of the East Bank in the Jordan Basin were made sufficient to irrigate 334,898 du, with water diversion requirements of 479 mcm/y. Of this, 175 mcm/y would come from side wadis flowing westward in Jordan to feed the Jordan River, 8 mcm/y from groundwater wells, and 296 mcm/y from the Yarmouk. No water was allotted to the East Bank from the Jordan River course, as the flows of the feeder tributaries in the amount of 175 mcm plus a share of the Yarmouk, along with the groundwater estimate, were sufficient to irrigate the arable lands in the Jordan Valley–East Bank. But water from the Jordan River was allotted to the West Bank, to be drawn from Lake Tiberias via the East Ghor Canal on the East Bank, and then delivered to the West Bank via a siphon under the Jordan River at Deir Alla.[25]

At the time of negotiations, because of increasing Syrian abstraction from water sources in the Yarmouk catchment, the average annual discharge of the Yarmouk at Adassiyya measured only 270 mcm/y on average, compared with the 402 mcm/y calculated in the Unified Plan (377 mcm for the East Bank and West Bank, and 25 mcm for Israel). The difference, 132 mcm/y on average, plus the amount of 14 mcm allocated to evaporation, apparently had been used by Syria upstream. Thus the total Syrian uses amounted to an average of 236 mcm/y, compared with its share of 90 mcm/y, an overrun of 146 mcm. Of this overrun, 65 mcm belonged to Jordan, as per Johnston's formula, and 81 mcm to the West Bank. It is believed that in 2005, Syrian uses of the Yarmouk sources were even greater.

The bilateral peace negotiations were a mechanism for the resolution of the water conflict between Jordan and Israel, not between Jordan and Syria. In abstracting from the Yarmouk, Israel's overrun was between 50 and 65 mcm, most of it in the winter months.[26] The task of the Jordanian negotiator was, as a minimum position, to confine Israel's share in the Yarmouk water to 25 mcm, as stipulated in the Unified Plan, and schedule the withdrawal of this quantity.[27] The Jordan River sharing was not on the table, because the Hashemite Kingdom's share in that river as specified in the Unified Plan was destined to the West Bank.[28] Other Arab shares in the Jordan north of the lake belonged to Lebanon and Syria.

Water obviously ranked high on the agenda of Jordan's negotiations with Israel, and the two countries came to an agreement over it by the early morning of Monday, October 17, 1994. Article 6 of the treaty addressed water and stressed the need for bilateral cooperation to alleviate the water shortage in each country. Cooperation was envisaged in the following fields, as quoted from Article 6 of the Jordan–Israeli Peace Treaty of October 26, 1994:

a. Development of existing and new water resources, increasing water availability, including cooperation on a regional basis, as appropriate, and minimizing wastage of water resources through the chain of their uses,
b. Prevention of contamination of water resources,
c. Mutual assistance in the alleviation of water shortages,
d. Transfer of information and joint research and development in water related subjects, and review of the potentials for enhancement of water resources development and use.[29]

Article 6 also refers to an annex of the treaty that details the water agreement. As specified in Annex II: Water Related Matters, the bilateral negotiations with Israel culminated in the following:

• Limiting Israel's share in the Yarmouk to 25 mcm/y, 12 mcm of which are due in the summer months (mid-May to mid-October) and 13 mcm during the rest of the year, a result commensurate with the Johnston Plan, which Jordan adopted as a basis of its water-sharing policy.

- A concession by Jordan to Israel to pump an additional 20 mcm of winter water that Jordan could not use for Israel's use, in return for delivery to Jordan of an equal quantity in the summer months. This would be virtual storage by Israel of Yarmouk floods in the amount of 20 mcm for Jordan's benefit—part of a provision in the Johnston Plan for the storage of Yarmouk waters.
- An additional quantity of 50 mcm of water fit for use for drinking purposes, the source of which was not defined.[30] Half of this quantity was being delivered from Lake Tiberias at a constant rate over the year until a desalination plant could be built to supply the entire quantity.[31] This quantity was over and above what Johnston envisaged as a share for Jordan.
- Desalinated water in the amount of 10 mcm/y. This quantity was to be supplied from Lake Tiberias until the desalinated water became available. Although not explicitly stated, this amount was in exchange for lower-quality Jordanian water in Wadi Araba that Israel was allowed to pump, as stated below.
- An increase of 10 mcm in uses by Israel from wells it drilled in Wadi Araba while a portion was under Israeli occupation (it since was returned to Jordan under the treaty). The quality of that water is marginal, at about 1,500 ppm total dissolved solids (TDS).

Other provisions were stipulated in Annex II addressing the protection of water quality and water structures in the territories of jurisdiction of each party. The Joint Water Committee was established for mutual cooperation and consideration of problems in implementation.

An improvement of the provisions of the Water Annex took place on March 10, 1998, when the Jordanian minister of water and irrigation and the Israeli minister of national infrastructure agreed to divert 60 mcm of Yarmouk floods to be stored in Lake Tiberias for Jordan's benefit. This was in line with the Johnston Plan, which stipulated a delayed implementation of such diversion five years after the implementation of the other components of the plan. The ministers also agreed to start building a diversion weir on the river at Adassiyya. The weir was completed in the year 2000, but diversion to Tiberias was not implemented.

The above shows that Jordan has achieved its objectives in limiting Israel's share in the Yarmouk to 25 mcm/y, as the Unified Plan stipulated, and succeeded in scheduling delivery of Israel's share in the Yarmouk. Furthermore, Jordan obtained more water from Israel than the Unified Plan specified.

No major problems have been reported in the work of the Joint Water Committee, and the water flow has been maintained across the borders since July 5, 1995. The water flow from Israel to Jordan proved to be crucial and extremely helpful, because it was used to keep the supply rate to Amman from the East Ghor Canal (renamed King Abdallah Canal in 1987) at its full capacity of 1.43 cms. The "peace water" was particularly helpful in the dry months. Still to be finalized are the erection of a desalination plant to supply Jordan with 60 mcm/y and clarification of the concessions of 20 mcm from the Yarmouk to Israel and from the lake to Jordan.

Policy in the Multilateral Working Group on Water Resources

Jordan's negotiation policy as part of the Water Resources Working Group of the multilateral conference was to provide its vision of cooperation in water resources after peace was achieved. Care was taken not to initiate any cooperative effort with Israel prematurely, because that would have hampered the progress of bilateral negotiations. Jordan presented its vision to cooperate in improving practices of water management and making more water resources available not only to Jordan and Israel, but also to the rest of the water-poor countries east of the Suez. It further proposed the establishment of a regional water databank accessible to all subscribers, suggested the initiation of pilot projects for data collection on desalination of seawater in Gaza and brackish water in Jordan, and endorsed the idea forwarded by Oman to establish a regional center for research on seawater desalination. Jordan's delegation aborted all attempts, before peace was reached, to initiate limited regional cooperation among Jordan, Israel, and the Palestinians, because such cooperation would have been premature and unproductive.

The multilateral conference and the work of all its groups stalled in 1996, when the Likud in Israel assumed power under Prime Minister Benjamin Netanyahu, and have not yet been restarted. The stagnation in the Middle East peace efforts undoubtedly impacts Jordan's cooperation with Israel. Perhaps the rather smooth management of the Water Annex in Jordan's peace treaty with Israel is an exception. Water-related matters have been progressing with the least difficulties. It is hoped that the peace process will regain momentum to achieve its goals of a just, lasting, and comprehensive peace in the region, which will promote regional cooperation.

Summary

Competition for water resources commenced in the early twentieth century, after the World Zionist Organization (WZO) decided to seek a national home for the Jews in Palestine. The competition intensified in the late 1930s. The WZO and the Emirate of Transjordan prepared competing water plans for the utilization of the Jordan River waters. The competition transformed into an international conflict after the proclamation of the Jewish state of Israel in 1948. Jordan had to rely on foreign aid both financially and technically, and Israel did the same.

Jordan's efforts to utilize the waters of the Yarmouk in 1952 were stalled by protests filed with the United States by the lower riparian on the river, Israel. The United States, eager to maintain stability in the Middle East and prompted by military clashes over water and other issues, dispatched a presidential envoy, Eric Johnston, in October 1953 to work out a plan acceptable to all the riparian parties. Johnston's efforts culminated, by October 1955, in the formulation of a Unified Plan for the development of the Jordan Valley, endorsed at the technical level by both Israel and a committee formed by the League of Arab States.

The flow of the Upper Jordan north of Lake Tiberias was estimated in the plan to be 1,031 mcm for the Jordan as it discharged into Tiberias, including a ceiling of 15 mcm/y of brackish water. The shares on the Upper Jordan were 35 mcm for Lebanon, 42 mcm for Syria, and 100 mcm for the West Bank (then part of Jordan), and the residue, about 554 mcm, was allocated to Israel. Evaporation accounted for the balance of the flow, or 300 mcm. The Yarmouk waters were shared in the plan as follows: 90 mcm for Syria, 25 mcm for Israel, and the residue, or 377 mcm, for Jordan (including the West Bank). Evaporation accounted for 14 mcm of the Yarmouk flow.

The Unified (Johnston) Plan served as a cornerstone in the policy of the United States concerning the efforts undertaken by any riparian party to develop and use any portion of the Jordan waters. The plan soon became Jordan's policy in relation to waters of the Jordan Basin. An international water agreement concluded between Jordan and Syria in 1953 for the utilization of the Yarmouk was compatible with the subsequent Unified Plan.

With commitments issued by Jordan to abide by its shares in the Unified Plan, the United States advanced grants to implement the largest irrigation project ever in Jordan, the East Ghor Canal Project. The project led to policies regarding land reform being legislated and implemented. Almost simultaneously with Jordan's project, the United States also supported through grants an Israeli project, Tiberias–Beit Shean, which would draw water from Lake Tiberias (Jordan River waters) to irrigate the Jordan Valley west of the river inside Israel.

Israel's attempts to divert portions of the Jordan waters to the Negev through its National Water Carrier triggered an Arab diversion of the upper tributaries. The June war of 1967 stopped the Arab project.

Jordan pursued a policy of regional cooperation with Syria and with Iraq to secure water needed for municipal purposes. Transfer of water from the Euphrates was found to be too expensive, but plans to build a dam on the Yarmouk as per an agreement with Syria finally succeeded. Regional cooperation with funding agencies has been fruitful. Cooperation with Iraq over water resources led to the training of Iraqi water professionals after the U.S.-led occupation of Iraq.

In bilateral negotiations as part of the Middle East peace process, the water conflict with Israel was resolved along the lines of the Unified Plan, with updates to Jordan's advantage. Jordan had to abide by the decision to disengage with the West Bank and negotiated its own water share with Israel, leaving the Palestinians to negotiate the share of the West Bank.

References

Blass, S. 1973. *Mei Meriva u-Ma'as* (Water in Strife and Action). Israel: Massada Press, Ltd. (in Hebrew). Cited in M. Lowi. 1993. *Water and Power.* United Kingdom: Cambridge University Press.

Haddadin, M. 2001. *Diplomacy on the Jordan: International Conflict and Negotiated Resolution.* Norwell, MA: Kluwer Academic Publishers.

Hays, J.B. 1948. *TVA on the Jordan: Proposals for Irrigation and Hydro-Electric Development in Palestine.* Washington, DC: Public Affairs Press.

Ionides, M.G. 1939. *Report on the Water Resources of Transjordan and Their Development.* London: Crown Agents for the Colonies. Published on behalf of the government of Transjordan.

Kerr, M. 1967. *The Arab Cold War, 1958–1967.* New York: Oxford University Press.

Lowdermilk, W.C. 1944. *Palestine, Land of Promise.* New York: Harper and Brothers Publishers.

MacDonald and Partners. 1951. *Report on the Proposed Extension of Irrigation in the Jordan Valley.* London, England: Sir Murdoch MacDonald and Partners.

Shlaim, A. 2000. *The Iron Wall: Israel and the Arab World.* New York: W.W. Norton & Company.

Stevens, G. 1956. *The Jordan River Valley.* International Conciliation series, no. 506. New York: Carnegie Endowment for International Peace.

———. 1965. *Jordan River Partition.* Stanford, CA: Hoover Institution.

Welling, T. 1952. Statement on the Yarmouk-Jordan Project. Paper presented before the Jordan Development Board on July 1, and cited in Haddadin 2001, 28.

Notes

1. Shlaim, an Israeli historian, reports that at the beginning of the war, the total number of all Arab troops was less than 25,000, and the Israelis fielded 35,000. By December 1948, the Israeli Defense Forces numbered 96,441 soldiers, but the Arab states could not match this rate of increase.

2. See the report of Abraham Bourcart to the Reverend John Wilkinson dated February 18, 1901 (Haddadin 2001, 446–52).

3. The government of the Hashemite Kingdom of Jordan commissioned a British

engineering consulting firm, Sir Murdoch MacDonald and Partners, in 1949 to appraise the potentials for the use of the Jordan River waters. The consultants made use of Ionides's plans, published in 1939. Israel published its seven-year plan in 1953, based on the work of Lowdermilk (1944; as detailed by Hays 1948), as well as those of Israeli engineers working for Mekorot, the Israeli water company, most notable of whom was Simcha Blass.

4. Palestinian refugees settled in Waqqas, Kreimeh, Khazma, Karama, and Rawda of the east Jordan Valley.

5. The Jordanian funds were made available through a loan extended to the government by the government of Great Britain.

6. The firm was Harza Engineering Company of 106 South Wacker Drive, Chicago, Illinois, which moved in the late 1980s to the Sears Tower, across Wacker Drive.

7. "International negotiations" was meant to imply negotiations between Jordan and Israel.

8. The international boundary is the Yarmouk watercourse or the railroad, whichever is closer to Jordan. The railroad is a narrow-gauge junction from Deraa to Haifa of the Hijaz Railroad, which in turn is a north–south junction of the Orient Express built before World War I to link Berlin with Baghdad.

9. Stevens (1965) reported that the study sponsored by the Tennessee Valley Authority (TVA) was actually under the auspices of both the UNRWA and the United States.

10. The riparian parties at the time were four: Lebanon, Syria, the Hashemite Kingdom of Jordan (including the West Bank), and Israel. Pursuant to Jordan's disengagement decree of 1988, the West Bank, represented by the Palestinian Authority since 1994, became a fifth riparian party.

11. Stevens (1965) reported that the TVA–Main study was under the auspices of UNRWA and the United States. In line with this report, one cannot imagine that a U.S. government agency such as the TVA, whose chairman is appointed by the president of the United States, would undertake such a mission without government approval.

12. Johnston was born in Washington, D.C., in 1895 and raised since early childhood in Spokane, Washington. He attended the University of Washington Law School in Seattle, but his studies were interrupted by his draft to join World War I as a marine in China. He headed the U.S. Chamber of Commerce for three consecutive terms and served under Presidents Truman, Eisenhower, and Kennedy. He died at George Washington Hospital in Washington, D.C., in August 1963.

13. For a detailed account of the Johnston mission and the talks he conducted with the Arab and Israeli parties, see Haddadin 2001, *46–121.*

14. The updated percentage is the estimate made by the Syrian team in the Joint Jordanian Syrian Committee on the Yarmouk, expressed in several meetings when I was heading the Jordanian team in 1986 and 1987.

15. Lebanon's foreign minister Salim Lahoud asked U.S. ambassador to Lebanon Heath on April 28, 1956, to seek such commitment from Israel. Lahoud reportedly made a similar request of the UN secretary general when he visited Lebanon that year.

16. For details of that "war," see Kerr 1967.

17. The consulting firm was Harza Engineering Company of Chicago, Illinois, one of the members of the Joint Venture that did the studies for the Master Plan for the development of the Jordan Valley in 1953–1954. Harza was also the firm commissioned to do the studies of the Bunger (Maqarin) Dam on the Yarmouk in 1952. The company has been retained by Jordan for studies of the Yarmouk several times since then.

18. The Baghdad Pact was a military alliance headed by the United Kingdom, with Turkey, Iraq, Iran, and Pakistan as members; concluded in 1955, it was aimed at confining the Soviet Union on its southern flank. Iraq pulled out of the pact immediately after the military coup of 1958. The Suez Campaign was a military operation waged against Egypt on October 29, 1956, by the United Kingdom and France, in collaboration with Israel. Israel occupied the Sinai Peninsula, and troops from the United Kingdom and France parachuted and occupied certain locations in the Suez Canal Zone. The invaders had to withdraw by virtue of pressures from the United States and the Soviet Union.

19. See Haddadin 2001, *138*.

20. This number was the Israeli figure stipulated in their proposal for a Memorandum of Understanding with Ambassador Johnston dated July 5, 1955. Johnston's final figure for the Israeli share in the Yarmouk was 25 mcm/y, as per the text of the Unified Plan, distributed to the parties by the Johnston mission in January 1956. Johnston stood by his figure despite Israeli protest. This remained a point of contention between Jordan and Israel until it was resolved during the bilateral peace negotiations in 1994.

21. The director of operations of the East Ghor Canal thought that Jordan was entitled to two-thirds of the flow plus 1 cms and Israel was entitled to one-third of the flow minus 1 cms. He believed that a document specifying the formula existed in the files of the National Planning Council, which coordinated the relations with the donor agencies. The formula began to be followed in 1988, after I had resigned my post as president of the Jordan Valley Authority, and I found it in a 1963 report submitted to the State Department by Johnston's technical arm, Dr. Wayne Criddle.

22. For details see Haddadin 2001, *244–51*.

23. Several water organizations were at work in the 1970s and mid-1980s. The National Planning Council, the predecessor of the Ministry of Planning, commissioned Howard Humphrey and Partners, a British consulting firm at that time, to do the study.

24. Howard Humphrey and Partners were selected through competitive bidding invited by the National Planning Council in 1981.

25. This path for the water of the West Bank was viewed as necessary, as the West Bank was part of the kingdom, and direct delivery to it from Lake Tiberias would pass through Israeli territories. This would require cooperation with Israel, which was not entertained by Jordan or any other Arab country.

26. Israel's summer use from the Yarmouk was an average of 22 mcm between mid-May and mid-October of every year; the rest of its pumping was during the other seven months. It stored that water in Lake Tiberias for later beneficial uses.

27. This was the annual figure accepted by the Arab Technical Committee and by the

concerned Arab Foreign Ministers in a high-level meeting in Beirut. For details, see Haddadin 2001, *97–101.*

28. This point was particularly sensitive. King Hussein in June 1994 instructed me, as the Jordanian negotiator, to abide by the decision of disengagement with the West Bank and leave the share of the West Bank for the Palestinians to negotiate and recover from Israel.

29. Article 6 of the Jordan–Israeli Peace Treaty can be found in its entirety at http://www.kinghussein.gov.jo/books_articles.html. Also see Annex II of the peace treaty, Water Related Matters, at the same Website.

30. This quantity of additional water was not stipulated in the Johnston formula of water sharing. It was later agreed to be provided by a plant to desalinate brackish-water springs in Israel.

31. This arrangement was reached at a summit meeting in Aqaba on May 7, 1997, after a clash between me, as Jordan's minister of water, and my Israeli counterpart, Gen. Ariel Sharon.

11

Conclusions

Munther J. Haddadin

This chapter summarizes the main points made in the various chapters of the book. The chronology is not essentially the same as the sequence of chapters, but rather flags the points that may have been made in more than one chapter. It is substance oriented, with an attempt to parallel the sequence of the chapters in the book as much as possible.

Recommendations are made here and conclusions are drawn. The contents, however, are not considered sufficient to reflect all that has been discussed in the book, and reference to the detailed material in each chapter is mandatory to make use of the presented material.

On Resources, Population, and Cost Recovery

Jordan consists of 92% arid and semiarid territories where rainfall averages 8.2 bcm per year, of which about 80% is lost to evaporation. The remaining 20%, an average of 1,740 mcm, becomes green water in the form of soil water that supports rain-fed agriculture, including agricultural and grazing lands and forests, and blue water in the form of surface flows in streams and groundwater infiltration into the aquifers. The amounts of these resources are 866 mcm for green water, 602 mcm for surface flows, and 198 mcm for groundwater. These last two add up to 800 mcm of indigenous blue water.

Jordan's water resources have an international component related to surface inflows from the Yarmouk River and the sharing of the Jordan River south of Lake Tiberias. Groundwater inflows from Syria contribute to the annual amount of blue-water resources. The international components are 284 mcm for surface flows and 68 mcm for groundwater flows, with international flows totaling 352 mcm. Total blue-water resources amount to 886 mcm of surface resources, of which 838 mcm can be economically developed, and 266 mcm of groundwater resources, making 1,104 mcm of accessible blue water.

The recycling of water adds about 80 mcm of gray water (treated wastewater) and 70 mcm of leakage and irrigation return flows, totaling 150 mcm/y in 2002. The grand total of all resources, including green water (866 mcm) and the said level of gray water and return flows, amounts to 2,120 mcm. With a population of 5.35 million people, the per capita share of the renewable resources potential becomes 396 m^3/y, or about 23.3% of the needs of a Jordanian estimated at 1,700 m^3/y. This relationship displays a substantial imbalance in the population–water resources equation, which manifests in a water stress of 2,525 persons per unit flow of water (1 million m^3/y), as compared with a balanced stress of 588 persons per unit flow.

The population–resources relationship will get even tighter with time as population grows and water stress further increases. This calls for unusual care in water management and close looks at the water policy as the stress escalates. A dual approach of demand and supply management appears to be the answer, but it is not the solution. Water tariffs need to be revisited periodically, particularly as the gross domestic product (GDP) grows. Tariffs should be so structured as to recover the highest possible percentage of the cost of service. Public education campaigns should be sustained, and water awareness courses should be considered for inclusion in the cycles of basic education. Stakeholder participation is a basic aspect of water governance, and it should be enhanced and enlarged. Beneficiaries have to have a say in the governance of their water resources and the quality of services received.

The quantification of green water (soil water) as presented in Chapter 1 is a welcome addition to the means of assessing water resources in Jordan and other countries. Green water can be listed as a renewable water resource of the country, and a better picture of the resource availability can be made. On the supply management side, it is noted that the amount of potential green-water resources that support rain-fed agriculture is as high as the amount of indigenous surface-water resources. Jordan has not paid as good attention to the management of green-water resources as it has to the blue-water resources. A shift in policy is warranted to accord green water the importance it deserves. Perhaps merging the production activities of the Ministry of Agriculture and the Ministry of Water and Irrigation to form a Ministry of Water and Agriculture may be a good idea. The marketing aspect of agriculture can be vested in an autonomous organization that would care for agricultural exports and imports and in the local agricultural markets; the marketing organization would coordinate with the Ministry of Water and Agriculture and the Ministry of Trade and Industry. Stakeholders, represented by the Farmers Association, must have a big stake in the marketing organization.

Conservation of water resources dictates that the resources at hand be protected against quality degradation and depletion. This calls for more stringent control over the abstraction of groundwater resources. Although the government moved in 2002 to legislate for such controls, that legislation has to be revisited to amend its articles and make them compatible with the cultural tradition and in line with the spirit of the Constitution. Conservation also dictates that green-water resources be preserved and put to efficient use. Urban encroachment on agricultural land is tantamount to green-water losses. Urbanization in the recharge areas of aquifers tilts the blue-water balance away from aquifer recharge toward surface runoff, a shift that entails some sort of blue-water reallocation to beneficiaries. Leaving rain-fed lands fallow is indicative of low efficiency of water resource uses. Legislation has to be enacted to preserve such resources and allocate nonarable lands to the construction of urban areas. Strong liaison between the Ministry of Water and Agriculture and the Ministry of Municipal Affairs should be assured and sustained to achieve this policy objective.

The contribution of leakage water to aquifer recharge should be reduced. Leakage from municipal water networks wastes energy and does not preserve water. Energy is wasted to lift the water from the aquifer and pump it into the networks only to have it leak back again to the same aquifer. Other expenses are incurred to treat the raw water and manage its service. Leakage therefore should be kept to a minimum, preferably less than 15%. Contribution of irrigation return flow, though quantitatively favorable, is not favored on environmental grounds. Irrigated farming in the recharge areas of aquifers poses a built-in disadvantage: farming inputs and dissolved solids are carried by irrigation water or rainfall down to the water table, polluting the aquifer and increasing its salinity. Most disadvantageous are pesticide residues and other chemical fertilizers. Therefore, a policy shift is warranted to discourage irrigated farming in the recharge areas of groundwater aquifers.

The role of wastewater in augmenting water resources has been practiced since the late 1960s. Wastewater in Jordan cannot be managed as waste. It has to be treated in plants for environmental, public health, and resource-related reasons. The quality of effluent from wastewater treatment plants should not be allowed below the standards given in the national specifications governing the process. For nearly a decade and a half, the major wastewater treatment plant at As-Samra has been operated above design capacity, at a time when the effluent is allocated to irrigation in the Jordan Valley after being blended with storm water in the King Talal Dam reservoir. The overuse has adversely affected the environmental and economic conditions of the irrigated areas dependent on the release from the dam. Care should be taken to continuously monitor the effluent of all the wastewater treatment plants and assure that it conforms to the national specifications. An update of the Wastewater Management Policy of 1998 is warranted to emphasize the increasing role of treated wastewater in augmenting the renewable water resources and allowing possible reallocation of fresh water currently used in irrigation. The policy update should be followed by legislation translating it into implementation.

Silver water, or desalinated water, is being generated in Jordan by the efforts of the Ministry of Water and Irrigation in the direction of supply management. Rightly, silver-water generation is done using brackish water of salinities much less than that of seawater. Brackish groundwater emerging as springs is relatively plentiful around the Dead Sea and in the Jordan Rift Valley at large. The drawback in using brackish water from the Rift Valley lies in the high elevations to which the finished water is to be pumped. This is energy intensive not only in overcoming the total dynamic head, but also in operating the reverse osmosis processes employed in the generation of silver water. The cost of energy is an exogenous factor over which government has no control. The total cost of silver-water service includes the distribution cost, in which a high percentage of nonrevenue water exists. Thus energy cost plus nonrevenue water cost, along with the high capital cost, constitute prohibiting factors in achieving full cost recovery before the GDP is boosted to higher levels. Instead of calling on Rift Valley resources for supply management, reliance on the sandstone aquifer underlying the territories of the entire country should be assessed and investigated. Although static lift from the sandstone may be similar to those of the Rift Valley, the capital cost and overall operating and maintenance costs may well prove to be more economical than in the Rift Valley case.

For the above and purely management efficiency reasons, nonrevenue water has to be brought under control. Rehabilitation and replacement of aging water networks and reorganization of operation systems to assure gravity flow service to households from storage reservoirs will reduce nonrevenue water but not eliminate it. Elimination of illegal connections to networks, as well as human resources training and incentives for excellence, will further reduce the percentage of nonrevenue water. Although the reduction of illegally obtained water will not affect the resource availability to households, it will improve the service through better percentages of cost recovery and improve equity in water tariff so that law-abiding beneficiaries do not have to pay for water illegally obtained from the network by others. The Ministry of Water and Irrigation has achieved good progress in network renewal and operation systems, but illegal connections to networks are still a standing problem. Nonrevenue water use in irrigation has been drastically reduced through sustained checks.

One factor remains overlooked in the charges for water and wastewater systems that is worthy of consideration for capital cost recovery. The percentage of vacant lots in urban areas is high. These are land plots earmarked for various uses in the zoning of cities, especially Amman. Water distribution networks have been installed and cover practically all of these empty lots. While municipal water (and wastewater) tariff is a function of water consumption, the empty land plots do not account for any water consumption or wastewater collection and treatment. No water and wastewater charges are currently levied from the owners of the empty lots. The capital cost of networks is undoubtedly sensitive to the length of the networks that are capable of serving the empty plots, but those have no need to connect to the service until they are built up. A capital (one-time) charge on each empty lot to recover the cost applied to it in the networks is fair. The installation of networks raises the price of empty plots, as do other factors of development. A policy adjustment aimed at charging the owners of empty land plots in cities and towns is warranted, and this should be legislated as well. This will improve, in a fair and constitutional way, the recovery of the capital cost of water supply and sanitation projects.

Full cost recovery of water services in Jordan may be elusive; the influx of people in various categories—refugees, displaced, returnees, foreign labor, and Iraqis seeking safety and security—has boosted the population to unprecedented levels. These waves of people live mostly in urban areas or adjacent camps. They have created accelerated stress on local water resources, and these areas have had to turn to remote resources—some, such as Azraq, were already committed to agricultural and environmental uses. The cost of water service continues to rise as the influx of people increases. The GDP has not paralleled the population growth and cost of living. Cross-subsidies in water and wastewater tariffs do not solve the cost recovery issues; rather, they help moderate the effect of distortion in income distribution. The poor thus are enabled to pay for their modest consumption of water, but the overall cost recovery is a function of the GDP itself. Unless the GDP is accelerated in growth, the full cost recovery of water and wastewater services may never be achieved.

In this regard, the Kingdom of Jordan should perform a detailed assessment of the costs it has incurred in housing the waves of people it has received since 1948, along with the natural growth this has caused, including the increased cost of providing utilities. It should seek compensation for its expenses when the issues of refugees are addressed in the context of Middle Eastern peace.

On Environmental Concerns

The water stress in any locality can be considered a measure of environmental stress as well. Total water stress of 2,525 persons per cubic meter compared with 588 persons per unit flow at equilibrium reveals how much Jordan's overall environment is stressed. The fact that a small percentage of the country (6.9%) receives more than 300 mm of rainfall, 12.5% between 100 and 200 mm, and 80.6% less than 100 mm testifies to the fragility of the environment. Fragile environments persist in arid and semiarid regions, and Jordan is no exception. Land use in Jordan is zoned as follows: grazing areas, 93.3% of total area; buildings and public facilities, 1.89%; forests and reforestation areas, 1.5%; water surface, 0.62%; and agricultural land, 2.69%. The large proportion of grazing land testifies to the fragility of the environment.

Environmental awareness in Jordan was birthed in the 1980s; up until then, the environment received little consideration in economic development planning. The first environmental assessment conducted for an economic project was done for the second stage of the Jordan Valley Project, including the Maqarin Dam and associated irrigation networks. The study was required and financed by USAID. Only donor pressure helped bring environmental issues into the spotlight, necessitating assessments of adverse environmental impacts of economic development projects and ways of mitigating them.

The diverse ecosystems in Jordan call for more sophisticated management of the environment. Some locations are Ramsar sites, such as the Azraq Oasis, an important station for migratory birds. Water reservoirs created by the construction of dams are also emerging as stations for migratory and local birds. More important among them are the Wadi Arab Dam in the north Jordan Valley and the Karama Dam in the middle Jordan Valley. Although the operation rules for both dams have been determined and geared for releases for irrigation and municipal uses from the Wadi Arab and irrigation uses from the Karama, the roles of both dams should be reconsidered, with an attempt made to factor in the environmental needs from each. Plans are being finalized (2006) to establish a tourism village around Karama Dam, but a revision of the operation rules has yet to be done to guarantee a basic environmental role for the reservoir.

The introduction in the mid-1950s of drilling rigs for groundwater investigation and development has accelerated the use of groundwater for irrigation and municipal uses. In many cases, the abstraction from aquifers through wells dried up freshwater springs that historically had emerged from these aquifers. The condition of the Amman-Zarqa Reservoir is a case in point. Here overabstraction from the aquifer dried up the spring around which Amman was resurrected starting in the 1870s, as well as the rest of the springs emerging from that reservoir that fed the base flow of the Zarqa River. Water abstraction from other aquifers similarly dried up springs such as the Wadi Shueib, Wadi Arab, and most of the side wadis flowing toward the Jordan River. Concerning the international water, overabstraction from the Yarmouk aquifer inside Syrian territories decreased Jordan's share in the river and consequently stressed irrigated agriculture in the Jordan Valley.

In several of the side wadis, the depleted springwater flows have been replaced with treated wastewater. Although this is a welcome replacement quantitatively, qualitative monitoring should occur. The environmental impact of treated wastewater should always be a concern. The changes in the pH values of irrigation water, the fitness of treated wastewater for use in dissolving solid farm inputs, and the impact on wildlife on the farms should be fully studied, and mitigation measures should be defined.

Jordan is a party to many international environmental conventions. The endorsement of these conventions facilitated legislation for environmental management, such as the National Environmental Strategy and its Action Plan, the National Agenda 21 document, and the National Biodiversity Strategy and Action Plan, among others. It is important to recognize the dynamic nature of the environmental situation in the country, and thereby the importance of updating these strategies and action plans. Management of the environment has left a lot to be desired, and it is hoped that the newly established Ministry of Environment will be able to exercise its mandate as stipulated by law. Environmental initiatives should be homegrown, with assistance from the outside where and when needed. Coordination and liaison with the Ministry of Water and Irrigation and the Ministry of Agriculture should be maintained at all times to ensure applicability of the relevant specifications and guidelines.

Point-source pollutants should be scrutinized, and any licenses issued for the initiation and operation of such sources reviewed and updated with the aim of protecting the environment. Wastewater collection networks and treatment plants should be expanded to progressively eliminate the need for local disposal systems on the premises. Permits issued to gas stations, industries, and other point-source pollutants have to be upgraded and existing ones revisited. Solid waste landfills should be assessed across the country and specifications set for their construction to radically reduce point-source pollution. Streets of cities and towns have to be kept clean to avoid contribution to pollution after the first rainfall.

Nonpoint-source pollution is harder to control. Irrigated farming is the greatest contributor to pollution. Perhaps legislation regulating the use of pesticides and other chemical inputs to farming, with strict enforcement, can alleviate the impact of such pollutants. Farming in the recharge areas of aquifers was addressed above.

The drying up of the Azraq Oasis should be arrested and its rehabilitation attempted. Recharge of the aquifer feeding it can be enhanced through artificial recharge, using water from surface streams and accumulated in mudflats. The role of nongovernmental organizations (NGOs) in watching for environmental degradation should be encouraged and enhanced.

Perhaps the most important environmental rehabilitation needed is that of the Lower Jordan River. The Unified Plan, serving as a basis for water sharing, has to be revisited. It calls for the full use of the water resources for agricultural purposes before any gets to the Lower Jordan. Environmental flows received no attention when the plan was drafted in 1953 and negotiated between 1953 and 1955. The environment of the flood plain of the river changed drastically as irrigated agriculture expanded. Mitigation measures have to be introduced through regional cooperation among the riparian parties. Rehabilitation of the river course received attention in the Jordan–Israeli Peace Treaty, but implementation of commitments made by both parties has yet to be honored.

The level of the Dead Sea has been receding with time as a result of human intervention. Riparian states diverted virtually all of the fresh water that used to discharge into the Dead Sea, and Jordan and Israel actively pursued the exploitation of the sea to extract minerals. Reduction of inflow and increased evaporation progressively lowered the surface level of the Dead Sea, from a historic −392 m to −418.30 m in 2006. The success of both Jordan and Israel in agreeing to a mitigation measure was a step in the right direction. A conduit transporting water from the Gulf of Aqaba to the Dead Sea has been agreed upon within the integrated development plan for the Jordan Rift Valley. Yet the progress has been sluggish in implementing such a grand scheme. The political obstacle of including the Palestinians was resolved in the spring of 2005, but the cooperative effort is moving slowly.

Environmental impact of the Red Sea–Dead Sea linkage, mixing water from the open Red Sea with that of the closed Dead Sea, which has a higher density and mineral concentration, needs to be fully studied and understood. The economic, social, and political impacts are positive, especially when the project is viewed from the angle of integrated development of the Rift Valley. Breaking the project down into public sector responsibilities and components that can be undertaken by the private sector has yet to be done, and financing sources still need to be identified and attracted.

On Legislation and Institutional Arrangements

Jordan's water legislation and the accompanying institutional arrangements date back to the 1930s and are varied and diversified. Water has always been identified in the legislation as a public property, but its management passed through municipal councils and central government organizations before the water responsibilities were vested in a single cabinet entity, the Ministry of Water and Irrigation. The Ministry of Health supervises the applicability of health standards to drinking water, and the Ministry of Environment oversees the environmental aspects. In 1996, through cooperation with the World Bank, Jordan initiated the option of management contracts, and it took a step further by adopting the option of forming water companies owned by public sector entities where the revenues of water and wastewater exceed the costs, as in Aqaba. More firsts in the country followed, primarily the build-operate-transfer (BOT) contracts, with government contributing to the financing portfolio.

The trend of involving the private sector is healthy and should be encouraged. Private sector management of water and wastewater is envisaged to entail higher efficiency and better quality of service at less cost. However, the interference of government in the water management process should be limited to supervising service quality, tariffs, and other public interest issues. Management affairs, personnel appointment, or termination of services by the private contracting party should not be subject to government interference. Investment in upgrading the water systems should be the responsibility of government. Contracted management should be expanded to cover the rest of the governorates outside Amman and Aqaba.

Green water is an integral part of the components of water resources and should receive attention to boost its efficiency of use. Legislation for the regulation of green-water exploitation as a private-public partnership was alluded to in the above discussion and should be pursued. Institutional arrangement to have it properly managed has been recommended above in the form of a unified Ministry of Water and Agriculture.

Finally, legislation for the management of Jordan's shares in international waters is needed. The implementation of the Water Annex in the Jordan–Israeli Peace Treaty, which was signed on October 26, 1994, and went into law on November 11, has to be examined and commitments honored. A misunderstanding exists over the application of measures agreed to regarding the Yarmouk. The annex stipulates that Jordan concede to Israel the pumping of winter water from the Yarmouk in the amount of 20 mcm over and above its share; it also stipulates that this concession is in return for Israel supplying Jordan with 20 mcm from Jordan River water immediately above the Dagania gates (Tiberias Reservoir). The text calls for "mutual concession" undertaken by both Jordan and Israel. In the application of this commitment, Israel has imposed the understanding that it will supply Jordan in summer with only the same quantity as it pumps from the Yarmouk in winter, which clearly is not what is meant by the text of the Water Annex.

Another commitment that has to be honored is the delivery to Jordan of the 50 mcm of water per year it is entitled to under Article 1c of the Water Annex. Only 25 mcm have been delivered annually since 1997. Additionally, the mutual commitment under Article 6 of the treaty to work together to alleviate water shortage has yet to be realized.

On Water Data Management

Jordan's water strategy adopted in 1997 stipulated the establishment of a comprehensive water databank, supported by a decision support system, a program of monitoring, and a system of data collection, entry, updating, processing, and dissemination of information. It was intended to become a terminal in a regional databank setup. By 2002, this policy on data management was enhanced by substantial implementation, and online telemetry systems were installed in 2003. Compared with the hard copy system of documentation and analysis, this step was both innovative and an eye opener, but it was not applied to all parts of the country. Its coverage of the entire country is desired and should be commenced as soon as possible. Hurdles in the way of total coverage lie in financial, technical, and human resource constraints.

Foreign aid has to be tapped to finance, preferably in grant form, the capital cost of system generalization. The sophisticated systems are expensive to install and require advanced training to operate and maintain, and the operating and maintenance costs are usually high too. Outside financial resources are not willing to advance grants for operation and maintenance, and government sources will have to be tapped. Because of their importance to sound resource development, management, and policy implementation, the recurring costs of advanced water data systems should receive priority in budgeting for the water sector.

Data collection has improved in the recent years, but data verification has not. Attention must be paid to the logistical needs of the Water Information Systems division and the financial needs of its operations. Standard operating procedures for data have to be prepared and adopted, and water quality assurance measures should be employed. Policies based on poor data will be as good as the data provided, and vice versa. Although data quality and use have been highly improved in the irrigation system of the Jordan Valley, data collection and quality have to be improved elsewhere, especially for wells supporting agriculture. In many cases, water data have been synthesized or estimated. This practice has to come to an end, and the legislation in force now helps attain this goal. Water meters should be installed on all wells abstracting groundwater, especially those supporting agriculture, and these wells should be checked periodically to ensure that the meters are working properly.

The introduction of the databank, and the subsequent improvement on it in the form of data diversity and its interactive mode with users, has been a leap in water data management. Still, more improvements are to come. Of particular importance are the issues of accessibility to the Water Information System (WIS) and publication of data. The attitude toward water data has to shift from treating it as confidential to honoring the right to know for all data seekers, especially researchers. A policy must be formulated to regulate this and specify the mechanics of accessibility and periodic publications. Interaction and cooperation with the World Meteorological Organization and the friendly donors will enhance the speed of implementation of the needed improvements.

On Development of Water Resources and Irrigation

The most important irrigation project ever undertaken by the government has been the Jordan Valley Irrigation Project. An average of 200 mcm of water, primarily from the Yarmouk, was put to use to produce off-season fruits and vegetables. Five dams were built to support irrigation in the valley between 1963 and 1977, three more were built between 1999 and 2003, and a major one is being built and due for completion in October 2006. Development of irrigated agriculture outside the Jordan Valley was done by the owners of the lots to be irrigated, and wells were drilled under license from the Water Authority of Jordan (WAJ).

Jordan's experience in the diversion of agricultural water to municipal uses dates back to the 1950s to meet the growing demand of the capital city of Amman. Interbasin transfers have also been practiced since the 1960s to provide for the needs of the northern governorate of Irbid from a groundwater aquifer used for irrigation. More expensive transfers were made in the 1970s and 1980s for municipal and industrial purposes in Amman, Zarqa, Aqaba, Irbid, and Karak. Municipal water shortages in Jordan triggered moves by the government toward regional cooperation in water resource transfers. Nationally, the progressive diversion of agricultural water to municipal needs naturally caused friction between the government departments in charge of each of these sectors. The demands by the environmental sector exacerbated the already complicated scene. Measures were taken to mitigate the adverse impacts on agriculture and the environment from the diversions of surface water. Adverse impacts on groundwater aquifers and springs resulting from overabstraction of groundwater have yet to be seriously addressed.

The behavior of neighboring riparian parties on the Yarmouk had a negative effect on Jordan's shares. Peace with Israel achieved the restoration of Jordan's water already used by Israel. More has to be done to arrest overabstraction by Syria from the basin and to recover Jordan's rightful shares of water from that nation.

Irrigation water efficiency witnessed progressive improvements since the conversion of distribution systems to pressure pipe networks in the late 1970s. Greater efficiency is to be expected in light of reductions in Jordan's water shares from the Yarmouk. On-farm systems, although highly improved over traditional irrigation methods, should incorporate technological advances. This calls for improvements in agricultural marketing, research and extension services, and credit facilities. To determine the efficiency of agricultural water use, one should look at the measurements of agricultural yields per unit flow of water, or tons per unit flow. To determine the economic efficiency, one should look at the financial yield per unit flow, the foreign exchange component of agricultural exports, and the number of jobs created by the water use, among other economically important indicators. To determine the strategic importance of water, one should measure the population of rural agricultural areas and look at the contribution water is making to the food supply.

Sustainability of irrigated agriculture should occupy a high rank on the national scale of priorities. Threats to it come from the diminishing flows of the Yarmouk, especially in the dry months; overabstraction from groundwater aquifers; and degradation of quality caused by point and nonpoint sources of pollution. Better monitoring of water quality should be employed and measures taken to eliminate causes of pollution, especially in the context of increasing reliance on treated wastewater reuse. More efforts have to be exerted to improve the efficiency of municipal water uses by reducing the nonrevenue water percentages in all the consumption centers in the kingdom.

Focus should be made on the efficiency of use of green (soil) water. Its equivalence with blue water indicates that green water is just as important quantitatively. It should be preserved through clear policy on urbanization, which has been eating up good agricultural land, the reservoir of green water. Perhaps better locations for urban areas, in light of the diversification of Jordan's economy, would be the Badia provinces. More research, extension service, and credit facilities should be accorded to rain-fed agriculture.

Greater attention has to be given to the skills of human resources working in the development of water resources. Training of personnel, continuing education courses, and training of "blue-collar" workers should be upgraded and condensed.

On Water Economics

Water, a social and economic good, does not meet free-market conditions in Jordan: a water market in Jordan with proper competition in buying and selling cannot be created, and costs and benefits are not exclusively private, especially when environmental concerns are considered. Irrigation water remains subsidized primarily because the importance to society of water in the hands of farmers is greater than the farmers' willingness to pay. With regard to municipal water, the poor cannot meet the cost of water, and subsidy is unavoidable.

In light of the economic restructuring program, other alternatives to water subsidies should be considered. As has been the case with bread since 1996, farmers can be subsidized according to agricultural output rather than water consumption, and poor people can receive cash subsidies in lieu of an across-the-board water subsidy.

The public-private partnership that has started with the management contract for Amman's water and wastewater services should be expanded to include other governorates of the country. BOT contracts can be let for water and wastewater projects where the government is involved as a partner in financing and a regulator to protect the public against overpricing, deterioration of service quality, and monopolies. The government should refrain from intervening in the works of the private partner in issues outside of these.

Water studies have focused on feasibility and left a lot to be desired concerning household demand and consumption. Simultaneously, studies on the household level should be done for the willingness to pay for water and wastewater services; virtually none have been done thus far.

Given the huge imbalance in the population–water resources equation, the government has to try harder to secure financing of its water package, with the year 2012 as a horizon. Delays in implementation for any reason will make the situation even tighter. Responses more likely will be at the expense of agriculture, in which case one impact—loss of production—is mitigated through food imports. Involvement of society via NGOs in the public campaign to manage demand is crucial and should be reinforced. Structuring water tariff schedules, with due consideration to equity, should be a standing issue on the agenda of the fast-moving ministers. Upward adjustment of water tariffs improved the efficiency of both irrigation water and municipal water supplies. In terms of cost recovery, irrigation water revenue per cubic meter in 2002, for example, averaged US$0.015 compared with a cost of US$0.050, ignoring scarcity rent, leaving a direct subsidy margin of US$0.035, or 70.7%. Better management of the subsidy is needed; it could be directed to production rather than water consumption. The parallel picture in the municipal and industrial sectors shows a cost of supply of US$0.11 compared with an average revenue of US$0.064/m^3, leaving a subsidy margin of US$0.048/m^3, or 43%. The operating and maintenance cost of municipal water and wastewater systems amounted to 2.5% of the GDP in 2003; Water Authority of Jordan (WAJ) revenue collected from subscribers amounted to 1.46% of the GDP, leaving an average of 1.04% subsidy. The beneficiaries' water and wastewater expenses amounted to 1.34% of the average per capita share of the GDP when the operating cost amounted to 2.29% of that share. When the capital cost is added, the water and wastewater services would claim a higher percentage of the beneficiaries' income.

Unaccounted-for, or nonrevenue, water imposes financial, economic, and social problems. The nonrevenue water percentage currently runs around 46%, but it has to be reduced to a reasonable limit of about 15%. The government did well in its efforts to have the old networks replaced, by which the nonrevenue water percentage dropped from 55% to 46%; more work is needed for better control of the networks.

Charges for overabstracted water from aquifers are likely to reduce the overabstraction but not to eliminate it. A different approach should be adopted, where overabstraction is considered an action against the law requiring punishment under the penal code. The current policy charges for—practically sells—the overabstraction, which means that if the price is right in connection with farm revenue, more abstraction is likely. This policy has to change to achieve the objective of eliminating and protecting the aquifers from illegal overabstraction.

The construction of dams enabled the transfer of water from spills in winter to beneficial use in the dry months. Six dams built in the Jordan River catchment enabled the increase of agricultural area by 145% and the overall profitability of farms there by 55%.

Efficient and socially acceptable use of scarce water resources can be achieved only if all the cost of use is met through pricing of the service or government subsidies; the latter would indicate that societal benefits from the water use exist beyond the private benefits. However, a water-for-free policy along with a run-down distribution system can often result in powerful and rich people getting water cheaply through self-financed pipe systems, while poor people buy water at excessive rates or drink unsafe water. A free-water policy has other problems as well. Longtime subsidization distorts people's perception of water as a scarce and therefore valuable resource. Low water prices are thus likely to engender excessive use and waste of water, thereby worsening an already tenuous situation. To reduce water and wastewater subsidies, part or all of the capital cost can be recovered through taxes on vacant lots that are part of the networks but do not use the systems. Capital cost is better recovered through measures other than the water tariff structure, such as water connection fees, land taxes, or other indirect measures.

A general increase in irrigation water price would overproportionally affect cultivation of vegetables, fruits, and citrus in the open field and support the tendency to change toward production in plastic houses, as these use water more efficiently and generate better farm income. Expected consequences would include a loss in the variety of cultivated crops and an increased requirement for investments by farmers. The latter might also engender further negative secondary effects, as small-scale family enterprises, which constitute the majority of farms in the Jordan Valley, might not be able to cope with the increased financial demands. Experience in the valley indicated that reducing water consumption by using both the pricing and quota instruments did not produce the desired results. Under a water quota per farm system, an increase in water price yielded more revenue but did not affect water use.

The Water Allocation System (WAS), an economic model by which government decisions can be guided, was developed recently. Constraints are built into the model whereby government policies can be reflected. The model treats water as an economic good with free-market effects, but the user sets constraints regarding social factors as enunciated by government policies. The use of WAS, especially when it is upgraded to be multiyear model, will be very helpful for governments and water planners.

On the Role of Trade

The modest per capita shares of blue water (206 m³/cap) and green water (162 m³/cap), augmented by gray water and return flows (28 m³/cap), can hardly match the water needs per capita in a country with a lower-middle income economy such as Jordan. Population demand for water generates water stress, measured as persons per unit flow, and that stress generates water strain in the country.

The water needs of a country depend on the standard of living, which is closely linked to the income category to which that country belongs. The needs to satisfy municipal, industrial, and agricultural uses are shown to be 1,200 m³/cap for the high income category, 1,500 m³/cap for the upper-middle, 1,700 m³/cap for the lower-middle, and 1,900 m³/cap/y for the low. Jordan belongs to the lower-middle income category, with such needs met at 1,700 m³/cap/y. If the supply in each category of use matches the demand, then the water strain vanishes (zero strain), and the country is said to be in a state of water equilibrium. The water stress at equilibrium is therefore 833 persons per unit flow in the high income category, 667 in the upper-middle, 588 in the lower-middle, and 500 in the low.

The blue-water equivalent of green water in Jordan is shown to be about 866 mcm/y on average. If this flow of blue water is used in irrigation, it will produce as much tonnage of agricultural yield as the green water generates. Clearly, the combined supply capacity of blue, green, and gray water can never match the aggregate water demand, thereby the high water stress, with 2,525 persons per unit flow compared with 588 persons at equilibrium, results in high water strain.

Water stress is relieved and water strain vanishes with imports of processed commodities and municipal water; conversely, stress and strain are exacerbated by commodity exports and foreign tourism. This is not to suggest that processed commodity exports and tourism are harmful; they definitely have their economic and social advantages, but these exports have a water price that should be factored in as officials examine the water situation and plans are drawn for tourism and exports of processed commodities.

Shadow water, or the water that is saved by importing commodities, plays a crucial role in the water equilibrium of Jordan. Crucial also is the maximization of the contribution of indigenous water resources to the country's water equilibrium, an objective that can be met by boosting the use efficiency and productivity of a unit flow of water resources with all its components. But no matter how high water use efficiency is boosted in Jordan, the role of shadow water will remain critical. As things stand today, Jordan's indigenous resources are only 23.3% of the needs for basic purposes in municipal, industrial, and agricultural uses. Without shadow water, the imbalance in the population–water resources equation would make life in the country unbearable because of water strain. Shadow water, crucial to Jordan as it is, is not without disadvantages (listed in Chapter 7 along with the advantages). However, it is important to keep in mind the high level of water stress in the country and how important shadow water is. Because shadow water enters the water equation through commodity imports, it becomes equally critical for the country to weave good trade relations with the rest of the world, particularly its trading partners, and maintain them in the best shape possible. It is interesting to note that the average cost of agricultural shadow water was about US$0.011, compared with the actual cost of irrigation water in the Jordan Valley of US$0.049. Preserving shadow water has financial advantages over the use of indigenous resources.

The ideological and political trends in the country should also be factored in. A highly water-strained country cannot afford to take up extremist ideologically motivated foreign policies. Rather, quiet diplomacy and a "middle-of-the-road" approach may be more appropriate for the situation. It is of utmost importance not to strain relations with the trading partners; the priority of good relations with them is as high as water security to the country. It is not surprising that some might suggest that a country should "put its mouth where its water is."

Important also is the raising of foreign exchange to service the foreign trade. Here commodity exports and tourism become a priority. Jordanian nationals working abroad provide a dual advantage to the country: the earnings of foreign currency they send back to the country, and the fact that their water consumption is charged to their host countries.

The water situation prompts Jordan to lead efforts of regional cooperation. Nearby countries in the region—Iraq, Syria, Lebanon, and Turkey—are blessed with water resources. Turkey in particular has an abundance of blue- and green-water resources, with plenty for export. Syria has an excellent potential in blue and green water. These facts are incentives to start a network of regional cooperation and establish joint trade boards to ease the stress on water resources that exists in practically all the countries of the Middle East and North Africa but the four named above.

On Water Governance

Water and wastewater management was initially the responsibility of municipal councils, aided where needed by central government. The institutional arrangements were modified, and central government assumed more responsibilities. Today the Ministry of Water and Irrigation is the government's arm in the administration and management of water and wastewater. Its four branches—the WAJ, Jordan Valley Authority (JVA), Ministry Secretariat, and Project Management Unit (PMU)—shoulder the responsibilities of management of municipal and industrial water and wastewater (WAJ), irrigation in the Jordan Valley (JVA), studies and contacts with donors (MWI Secretariat), and grand water and wastewater projects (PMU). The WAJ and PMU have their own Treasury, separate from the government Treasury and its rules, but the JVA and MWI Secretariat budgets are part of the government Treasury obligations. In essence, the PMU is an entity created to facilitate the implementation of large water and wastewater projects usually funded by friendly donors, and its creation was part of their requests and demands. It is carrying out responsibilities originally vested in the WAJ by law.

Participation of the private sector with these entities is built into the composition of the boards of directors of the WAJ and JVA. No representation of the beneficiaries or the private sector exists in either the MWI or PMU. However, today some NGOs care for water and the environment and are active in their pursuit of matters related to both fields. Universities and the Royal Scientific Society administer water surveillance activities through their staffs and laboratories. The creation of new water institutions has been the historical trend in Jordan, mostly instigated by donor agencies and the need for foreign assistance to implement sizable water projects. The propensity in Jordanian governments for fast turnover or reshuffle of the cabinets has been a persistent handicap that emerged from the formation of the portfolio of minister of water. The Ministry of Water and Irrigation is usually a victim of fast turnover, to the disadvantage of water administration.

The water sector is primarily a technical sector with immense social, economic, political, public health, and environmental impacts. Jordan's traditional donors have had continuous interest in the sector and its governance to ensure that the objectives for which they have advanced loans and grants are met, and that the interactions between them and the sector's officials are maintained on solid grounds. This, however, does not justify the frequent shifts in the management of the sector and its governance. These are concerns for the Jordanians to worry about and upgrade themselves.

The experience gained so far reinforces the notion that government administration of the water sector should be centralized in planning and implementation, and less so in operation and maintenance. This suggests the creation of one water authority, headed by an experienced and seasoned professional as its CEO, with two assistants, one for planning and implementation of the plans, and another for operation and maintenance of the water and wastewater systems. A second CEO would take charge of the agricultural components of water use, the chores currently carried out by the Ministry of Agriculture, and oversee environmental concerns. The two authorities would come under a minister of water and agriculture, who would head the two boards of the authorities. Each CEO would assign a slot for the participation of the NGOs and the reinforcement of their roles. An effort should be made to downsize the staffs of the constituents of the Ministry of Water and Irrigation and the Ministry of Agriculture. Special attention should be paid to incentives given for excellence and severe punishment imposed for corruption.

The involvement of beneficiaries in water management and decisions on matters affecting them should be enhanced and reinforced. Details currently carried out by the cadres of the Ministry of Water and Irrigation and Ministry of Agriculture are better done by the beneficiaries themselves or by the private sector. The formation of water users' associations, especially for irrigation operations, would be helpful. The organization of water and wastewater councils in the various districts (north, middle, and south) would also be helpful, not only for performing detailed work, but also for obtaining popular support for the desired water governance, including water tariff adjustments and water and wastewater means to recover capital costs.

The shift to management contracts for water and wastewater services is a welcome trend. It should be expanded to all governorates, and Jordanian firms should be set up for the purpose. These firms can form joint ventures with international firms to bid for contracts with the water authority. In the process, a phaseout of WAJ employees can be done as the able among them are recruited by the new firms. Assurance of quality of service should be a priority in these contracts. For capital projects, BOT contracts should always be considered, with the likelihood of government participating in project cost to ensure success. Capital cost recovery should be shifted away from the water bill and be met through assessment of water and wastewater capital payment by the owners of all lots in the service area, the vacant ones included. Full cost recovery should be achieved for operation and maintenance of water systems. Full or partial capital costs of future irrigation projects should be recovered from the beneficiaries, especially once the social gains from irrigation projects manifested by land redistribution have been realized. Cost recovery can be enhanced if marketing of agricultural products is improved, a task that it is suggested be entrusted to an agricultural marketing organization attached to the minister of trade and industry.

Finally, the control of groundwater abstraction cannot stay lukewarm; it has to be enforced with unwavering determination. The government should use every possible means to protect this valuable resource and assure its sustainability. The legislation of 2002 should be amended to cancel the overcharge for overabstraction and replace it with financial penalties and closure of the violating wells.

On Social and Cultural Impacts

Water plays an important role in the rituals of both Islam, the religion of the majority of Jordanians, and Christianity, to which a small portion adhere, and both shape the attitudes of their followers toward water. The notion held by some Western scholars that Islam considers water a free commodity is not correct. Rather, the hadith of the Prophet stipulates that water is free unless an effort is made to obtain (serve) it, in which case a price can be charged for its service. An Islamic fatwa approved the reuse of treated municipal wastewater, opening the way for this important water resource to be utilized. Jordan embarked on the reuse of treated wastewater in irrigation in the late 1960s.

The accessibility to water services, much improved after the 1977 UN Water Conference in Mar Del Plata, is clouded by the resource shortage. Water is rationed, and service is not continuous. Better efforts have to be made to ensure equity in water distribution. In fact, poorer neighborhoods should be accorded higher care so that no family finds itself forced to purchase drinking water from vendors. Such purchases by the poor tax their modest income because of the high cost compared with what this bracket of society pays for the network water. The rich can afford to purchase from vendors, but the poor cannot.

Likewise, in rationing irrigation water, care should be taken to serve the seasonal crops, usually owned by sharecroppers and owner–operators. Although constraints on the ground dictate that the higher capital investment be protected, equity dictates that excessive protection under conditions of scarcity be avoided, as it invariably harms the interests of the poor.

Allocation and reallocation of water resources should consider equity among the different user sectors. Environmental needs, long neglected, have to be factored in as well. The current uses of groundwater for agriculture in the Highlands should be assessed to study the economic feasibility, social appropriateness, and environmental impact of diverting such waters to municipal uses or for the sake of reducing the overabstraction from the aquifers.

Care should be taken in the process of diverting irrigation water from the Jordan Valley to municipal uses in the Highlands. The return wastewater should be so treated as to comply with Jordanian specifications and Food and Agriculture Organization and World Health Organization guidelines. Reuse of the effluent in irrigation should not be a liability for the agricultural produce in domestic or export markets. As the importance of treated wastewater increases with time, so should the importance of its supervision and surveillance.

The integrated development of the Jordan Valley, with water and irrigated agriculture as its backbone, saw economic, social, cultural and environmental successes, indicating that the approach should be used elsewhere in rural development. The area has seen substantial gains in farm incomes, per capita income, cultural attitudes and practice, and the role of women and their entry into the labor force of services. It has also realized drastic improvements in educational achievements, health care, child morbidity incidence and mortality rates, life expectancy at birth, the quality of life in general, and many other social and cultural indicators. Politically, the valley emerged strong on the political map of Jordan when several members of Parliament were elected and several of its natives occupied ministerial positions.

More recently, however, the relaxation of attention to integrated development has been causing setbacks in the valley's social, economic, and cultural spheres. Comprehensive studies are lacking since 1987, when one was produced by consultants to the JVA with the generous support of USAID. The government should reexamine the diluted emphasis on integrated development in the Jordan Valley to sustain the drive it had started and maintained since 1973.

Social equity issues have been addressed mostly through water tariff structure, as well as through administrative decisions. Social equity should be observed when further upward adjustments of water tariff are considered. Capital cost recovery should be set aside from the water bills, which should charge for operation and maintenance only. It would not be fair to have the water users pay for the capital of networks built to serve not only them, but also a host of vacant lots scattered throughout cities, towns, and villages.

The shrinking and removal of subsidies to agriculture reduce the dispensable income of farming families, making it more difficult for them to improve the lot of their children through education, training, and health care. Care should be exercised when decisions in that direction have to be taken.

On Management of International Water Shares

The Unified Plan for the Development of the Jordan Valley, worked out by Ambassador Johnston of the United States in 1955, formed the basis of Jordan's claim to water shares in the Yarmouk and Jordan rivers. After the disengagement decision with the West Bank, the share from the Jordan River at Lake Tiberias stipulated in the plan was dropped from Jordan and left for the representatives of the West Bank, the Palestine Liberation Organization, to claim and recover from Israel. Jordan upheld the claim to its shares in the Yarmouk River, on which Syria is an upstream and Israel is a downstream riparian party.

Syria drifted away from the course specified in the Unified Plan as far back as 1967 and has acquired a substantial amount of Jordan's share specified in the plan and in a bilateral treaty with Jordan concluded in 1953. Jordan entered into negotiations with Syria and had this treaty replaced with another in 1988, to the advantage of Syria. Still, the acquisition by Syria of Yarmouk Basin water exceeds even the second treaty.

Although Jordan's urgency to build a dam across the Yarmouk, and thereby entertain Syrian demands, is understandable, it is not understandable why Jordan should not proactively pursue the protection of the flow of springs below 250 m above sea level, assigned to it under both treaties. Syrian abstractions from groundwater in the basin have materially affected the flow of these springs, and the base flow of the Yarmouk in the dry months dwindled to a little stream. This has noticeable impacts on agriculture in the Jordan Valley, and Jordan should insist on revisiting the articles of the treaty for both protecting the flow of the lower springs against groundwater abstraction in Syria and pro rata and simultaneous filling of surface dams in Syria and the Wehda Dam in Jordan (due for completion in October 2006). Failing that, the losses in the Jordan Valley will be substantial.

Water relations with the downstream riparian party, Israel, are fine considering the tense political situation that has persisted since 1999. However, a determined pursuit by Jordan of the implementation of the treaty provisions is needed. The aim should be to secure the rest of the additional water stipulated in the Water Annex (Article 1.c) and correct the ongoing misinterpretation of the meaning of Article 1 concerning the Jordanian concession to Israel to pump 20 mcm of Yarmouk winter flow in return for Israel delivering to Jordan in the summer months 20 mcm of water. The ongoing interpretation says that Israel is "storing" winter water for Jordan's use in the dry months. Although this can be interpreted as "virtual storage," it is not a real one. What the text says is not storage, but a concession by Jordan. If sufficient flow is available for Israel below the diversion structure, Israel will be entitled to a ceiling of 20 mcm; if the availability is less than that, then Israel pumps what there is. The same interpretation applies to Israel's reciprocal supply of 20 mcm from the "Jordan River immediately upstream from the Dagania gates": if enough water is there to pump a ceiling of 20 mcm to Jordan, then Jordan gets the amount; if not, Jordan gets what there is.

Joint work by Jordan and Israel has to start to build the two dams on the Jordan River as stipulated in the Water Annex; an alternative would be to do away with the two dams but provide Jordan with a storage space in Lake Tiberias. This alternative solution was agreed upon in writing between Israel's minister of national infrastructure and Jordan's minister of water and irrigation on March 10, 1998. Copies of the agreement were deposited with each party.

Joint work by the two parties has yet to commence on the rehabilitation of the Jordan River below Lake Tiberias. Industrial pollutants and wastewater diverted to it by Israel should be stopped as a first step, and means to improve the quality of the remaining flow agreed upon between the two parties.

Groundwater investigation in Wadi Araba has to start jointly by the two parties to determine whether the hydrogeological conditions allow the abstraction of more groundwater to supply Israeli farms there in the amount of 10 mcm/y over the pre-1994 level of 6 mcm/y. No increase in groundwater abstraction should be allowed before examining the potential safe yield of the aquifers concerned.

Finally, a Palestinian share is due from the Yarmouk, calculated to be about 80 mcm under the assumptions made in the Unified Plan. Such a quantity plus a good portion of the Jordanian share under that plan are now being used by Syria. This situation requires a resolution through negotiations when the time comes for Israeli–Palestinian talks on final status and for Jordanian–Palestinian–Israeli–Syrian joint talks. Such a requirement is a reminder of the imperative of regional cooperation on the Jordan Basin, which should include Lebanon. Cooperation among all the riparian parties will result in gains for each as well as collectively.

Abbreviations and Acronyms

AGWD	Aqaba Governorate Water Directorate
AID	Agency for International Development
APC	Arab Potash Company
ASAL	Agricultural Sector Adjustment Loan
ASEZ	Aqaba Special Economic Zone
ASEZA	Aqaba Special Economic Zone Authority
atm	atmosphere
AWC	Aqaba Water Company
AWSA	Amman Water and Sewerage Authority
bcm	billion cubic meters
BGR	Federal Institute for Geosciences and Natural Resources of Germany
BOD	biological oxygen demand
BOT	build-operate-transfer
cap	capita
CBD	Convention for Biological Diversity
cms	cubic meters per second
CWA	Central Water Authority
CWR	crop-water requirements
DOS	Department of Statistics
du	dunum
DWSC	Domestic Water Supply Corporation
EC	electrical conductivity
EGCA	East Ghor Canal Authority
EIB	European Investment Bank
EU	European Union
FAO	Food and Agriculture Organization (United Nations)

FOG	fat, oil, and grease
GCEP	General Corporation for Environmental Protection
GDP	gross domestic product
GIS	geographical information system
GTZ	Gesellschaft für Technische Zusammenarbeit
ha	hectare
HCU	historic consumptive use
ICA	International Cooperation Agency (of the United States)
ICP	internal competitiveness program
IDA	International Development Association (World Bank)
IPC	Iraq Petroleum Company
JD	Jordan dinar
JICA	Japan International Cooperation Agency
JRTRC	Jordan River Tributaries Regional Corporation
JVA	Jordan Valley Authority
JVC	Jordan Valley Commission
KFW	Kreditanstalt für Wiederaufbau
km	kilometer
KTD	King Talal Dam
L	liter
μs	micro siemens
m	meter
M&I	municipal and industrial
mcm	million cubic meters
MENA	Middle East and North Africa (World Bank)
mm	millimeter
MOA	Ministry of Agriculture
MOE	Ministry of Environment
MOFA	Ministry of Foreign Affairs
MWI	Ministry of Water and Irrigation
$N\text{-}NO_3$	Nitrate
NBSAP	National Biodiversity Strategy and Action Plan
NEAP	National Environmental Action Program
NES	National Environmental Strategy
NGO	nongovernmental organization
NGWD	Northern Governorates Water Directorate
NWMP	National Water Master Planning
NPC	National Planning Council
NRA	Natural Resources Authority
NRW	nonrevenue water
O&M	operating and maintenance
OECF	Overseas Economic Cooperation Fund
PAH	polycyclic aromatic hydrocarbons

PLO	Palestine Liberation Organization
PMU	Project Management Unit of the Water Authority of Jordan
ppb	parts per billion
ppm	parts per million
Ramsar	International Convention to Conserve Wet Lands (named after the city in Iran where the convention was held)
RSCN	Royal Society for the Conservation of Nature
SCADA	System Control and Data Acquisition System
sec	second
TCA	Technical Cooperation Agency (of the United States)
tss	total soluble solids
TVA	Tennessee Valley Authority
UNDP	United Nations Development Programme
UNEP	United Nations Environment Programme
UNRWA	United Nations Relief and Works Agency for Palestine Refugees
UNTSO	United Nations Truce Supervision Organization
USAID	United States Agency for International Development
USDA	United States Department of Agriculture
WAJ	Water Authority of Jordan
WAS	Water Allocation System
WDM	water data management
WEPIA	Water Efficiency and Public Information for Action
WHO	World Health Organization
WIS	Water Information System
WMIS	Water Management Information System
WMO	World Meterological Organization
WTO	World Trade Organization
WZO	World Zionist Organization
y	year

Index